Studies in Computational Intelligence

Volume 905

Series Editor

Janusz Kacprzyk, Polish Academy of Sciences, Warsaw, Poland

The series "Studies in Computational Intelligence" (SCI) publishes new developments and advances in the various areas of computational intelligence—quickly and with a high quality. The intent is to cover the theory, applications, and design methods of computational intelligence, as embedded in the fields of engineering, computer science, physics and life sciences, as well as the methodologies behind them. The series contains monographs, lecture notes and edited volumes in computational intelligence spanning the areas of neural networks, connectionist systems, genetic algorithms, evolutionary computation, artificial intelligence, cellular automata, self-organizing systems, soft computing, fuzzy systems, and hybrid intelligent systems. Of particular value to both the contributors and the readership are the short publication timeframe and the world-wide distribution, which enable both wide and rapid dissemination of research output.

The books of this series are submitted to indexing to Web of Science, EI-Compendex, DBLP, SCOPUS, Google Scholar and Springerlink.

More information about this series at http://www.springer.com/series/7092

Takayuki Ito · Minjie Zhang ·
Reyhan Aydoğan
Editors

Advances in Automated Negotiations

Springer

Editors
Takayuki Ito
Graduate School of Engineering
Nagoya Institute of Technology
Nagoya, showa, Aichi, Japan

Minjie Zhang
Computer Science and Software Engineering
University of Wollongong
Wollongong, NSW, Australia

Reyhan Aydoğan
Özyeğin University
Istanbul, Turkey

ISSN 1860-949X ISSN 1860-9503 (electronic)
Studies in Computational Intelligence
ISBN 978-981-15-5868-9 ISBN 978-981-15-5869-6 (eBook)
https://doi.org/10.1007/978-981-15-5869-6

This Springer imprint is published by the registered company Springer Nature Singapore Pte Ltd.
The registered company address is: 152 Beach Road, #21-01/04 Gateway East, Singapore 189721, Singapore

Preface

This book includes selected revised and extended papers from the 11th International Workshop on Automated Negotiation (ACAN2018), which was held in Stockholm, Sweden, in July 2018. Automated negotiations have been widely studied and are one of the emerging areas of research because of the recent drastic advance of the AI systems. In the future society, there could be multiple AI systems which act in a society. They are designed by different companies and would have different optimization policies. In such situations, there could be several conflicts among AI systems. Then, the concepts on automated negotiations become the real matter. This book gathers the state-of-the-art research works on automated negotiations.

The complexity in an automated negotiation depends on several factors: the number of negotiated issues, dependencies between these issues, representation of the utility, negotiation procedural and protocol, negotiation form (bilateral or multiparty), time constraints negotiation goals, and so on. Complex automated negotiation scenarios are concerned with negotiation encounters where we may have, for instance, a large number of agents, a large number of issues with strong interdependencies, real-time constraints, concurrent and inter-depended negotiation, etc.

In order to provide solutions in such complex automated negotiation scenarios, research has focused on incorporating different technologies including search, CSP, graphical utility models, Bayesian nets, auctions, utility graphs, optimization, and predicting and learning methods. The applications of complex automated negotiations could include e-commerce tools, decision-making support tools, negotiation support tools, collaboration tools, as well as knowledge discovery and agent learning tools.

Researchers are actively exploring these issues from different research communities. They are, for instance, being studied in agent negotiation, multi-issue negotiations, auctions, mechanism design, electronic commerce, voting, secure protocols, matchmaking and brokering, argumentation, cooperation mechanisms, uncertainty modeling, distributed optimization, and decision-making and support systems, as well as their application areas.

This book consists of the following parts:

- Part I: *Modern Agreement Models and Mechanisms* and
- Part II: *Negotiating Agents Competition, Tools and Evaluation Metrics*

The chapter titled "Deniz: A Robust Bidding Strategy for Negotiation Support Systems" is the invited paper authored by Catholijn M. Jonker and Reyhan Aydoğan, who were one of the main founders on the automated negotiating competition.

Finally, we would like to extend our sincere thanks to all authors. This book would not have been possible without the valuable support and contributions of those who cooperated with us.

Nagoya, Japan
Istanbul, Turkey
Wollongong, Australia
October 2019

Takayuki Ito
Reyhan Aydoğan
Minjie Zhang

Contents

Contributors

Reyhan Aydoğan Department of Computer Science, Özyeğin University, Istanbul, Turkey;
Interactive Intelligence Group Delft University of Technology, Delft, The Netherlands

Tim Baarslag Centrum Wiskunde & Informatica (CWI), Amsterdam, The Netherlands

Umut Çakan Department of Computer Science, Özyeğin University, Istanbul, Turkey;
Interactive Intelligence Group Delft University of Technology, Delft, The Netherlands

Arnab Chatterjee TCS Research, New Delhi, India

Katsuhide Fujita Tokyo University of Agriculture and Technology, Tokyo, Japan

Takayuki Ito Nagoya Institute of Technology, Gokiso, Nagoya, Japan

Catholijn M. Jonker Interactive Intelligence Group Delft University of Technology, Delft, The Netherlands;
Leiden Institute for Advanced Computer Science, Leiden University, Leiden, The Netherlands

Ryohei Kawata Tokyo University of Agriculture and Technology, Koganei, Tokyo, Japan

Onur Keskin Department of Computer Science, Özyeğin University, Istanbul, Turkey;
Interactive Intelligence Group Delft University of Technology, Delft, The Netherlands

Zahra Khosravimehr Facuty of Computer Engineering, University of Isfahan, Isfahan, Iran

Faria Nassiri-Mofakham Facuty of Computer Engineering, University of Isfahan, Isfahan, Iran;
Intelligent and Autonomous Systems, Centrum Wiskunde & Informatica, Amsterdam, The Netherlands

Gautam Shroff TCS Research, New Delhi, India

Vishal Sunder TCS Research, New Delhi, India

Onuralp Ulusoy Department of Information and Computing Sciences, Utrecht University, Utrecht, The Netherlands

Lovekesh Vig TCS Research, New Delhi, India

Pınar Yolum Department of Information and Computing Sciences, Utrecht University, Utrecht, The Netherlands

Modern Agreement Models and Mechanisms

Let's Negotiate with Jennifer! Towards a Speech-Based Human-Robot Negotiation

Reyhan Aydoğan, Onur Keskin, and Umut Çakan

Abstract Social robots are becoming prevalent in our society, and most of the time, they need to interact with humans in order to accomplish their tasks. Negotiation is one of the inevitable processes they need to be involved to make joint decisions with their human counterparts when there is a conflict of interest between them. This paper pursues a novel approach for a humanoid robot to negotiate with humans efficiently via *speech*. In this work, we propose a speech-based negotiation protocol in which agents make their offers in a turn-taking fashion via speech. We present a variant of time-based concession bidding strategy for the humanoid robot and evaluated the performance of the robot against human counterpart in human-robot negotiation experiments.

1 Introduction

Nowadays, robots are becoming an intrinsic part of everyday life, and in the near future, we will presumably see them almost everywhere in our society (e.g., work, home, hospital, school) [1, 2]. Even now, robots get to work in classrooms, hotels, airports, and so on. As robots become more human-like, they will be integrated into the society more widely. To carry out our tasks, we may need to collaborate with them more often and even negotiate with them on task allocations or other daily issues. For this reason, there is an urgent need for designing negotiating agents that can interact with humans effectively [3].

R. Aydoğan (✉) · O. Keskin · U. Çakan
Department of Computer Science, Özyeğin University, Istanbul, Turkey
e-mail: reyhan.aydogan@ozyegin.edu.tr

O. Keskin
e-mail: onur.keskin@ozyegin.edu.tr

U. Çakan
e-mail: umut.cakan@ozyegin.edu.tr

Interactive Intelligence Group Delft University of Technology, Delft, The Netherlands

T. Ito et al. (eds.), *Advances in Automated Negotiations*, Studies in Computational Intelligence 905, https://doi.org/10.1007/978-981-15-5869-6_1

Although automated negotiation has been studied for several decades and there already exist a variety of negotiation frameworks [4–12], the nature of "human-agent negotiation" requires considering different dynamics [13]. For example, it is possible to make hundreds of offers to reach an agreement in automated negotiation; however, this is not feasible for human negotiator. Moreover, the way of communication is another issue to be taken into account in human-agent negotiation as the rules of interaction. Therefore, a number of studies have been carried out on designing protocols for human-agent negotiation. For instance, Avi et al. propose a chat-based negotiation framework supporting to some extent using natural language processing and issue-by-issue negotiation [14]. Mell and Gratch have introduced IAGO framework, which allows a human negotiator to exchange offers, emotions (via emoji), preference statement, and free chat [15]. Jonker et al. develop a negotiation support tool, Pocket Negotiator, which aims to help human negotiator by means of some analytics and recommendation [16].

From the point of view of how aforementioned text-based negotiation frameworks function in HRI, it would be more convenient and effective to communicate via speech. Furthermore, as the reciprocal interaction of engagement is an inevitable part of human communication [17], a turn-taking fashion interaction would be appropriate for human-robot negotiation (HRN). Accordingly, this paper introduces a speech-based negotiation protocol, in which a humanoid robot negotiates with a human counterpart via speech by means of speech recognition and text-to-speech methods. A Nao humanoid robot was used to represent a human negotiator the action it should take was commanded remotely in [18]. In our case, our robot makes all decisions by itself in multi-issue negotiation. In that sense, this is the first humanoid robot agent negotiating with human autonomously.

To date, a variety of negotiation strategies have been developed for automated negotiation [19–21]. In this work, we propose a novel negotiation protocol using a set of arguments. We present a variant of a time-based bidding tactic, which changes its behavior stochastically between *Conceder* and *Boulware* [22]. A carefully designed user experiment has been conducted to study the performance of the proposed negotiation strategy in human-robot negotiation.

The rest of the paper is organized as: Sect. 2 explains the proposed negotiation protocol and strategy elaborately while Sect. 3 presents our experimental design and empirical evaluations of our findings. A list of related work is given in Sect. 4. Lastly, we summarize our contributions and discuss future work in Sect. 5.

2 Proposed Negotiation Framework

In the proposed framework, a humanoid robot Nao namely Jennifer[1] negotiates with a human negotiator to come up with a mutual agreement. In the following part, proposed speech-based negotiation protocol governing the interaction between Jennifer and a human negotiator is explained.

[1] In this paper, "Jennifer" is used to refer our humanoid robot.

2.1 Speech-Based Negotiation Protocol

Communication medium (e.g., *speech*, *vision*, or *text*) plays a crucial role in human communication [2]. "How we say" is just as important as "what we say." Therefore, it is important to select the best way of communication while designing a negotiation protocol for HRN. Most existing work on human-agent negotiation has used a text-based communication; however, human-human negotiations are mostly carried out through speech. In addition to this, establishing a communication with a human through speech rather than using a text-based method is a more natural and fluent way. Therefore, we propose a **S**peech-based **H**uman-**R**obot **N**egotiation **P**rotocol, called **SHRNP**, which is a variant of Alternating Offers Protocol [23].

Figure 1 shows the FIPA representation of this protocol. According to this protocol, Jennifer initiates the negotiation by asking whether the human negotiator is ready to make an offer (Turn 1). Human negotiator should tell Jennifer that she/he is ready to make her/his offer (Turn 2). Note that she or he should say "ready" at this stage (e.g., "I am ready"). When Jennifer recognizes the word "ready," she says that she is ready to hear her/his offer (Turn 3). Then, the human negotiator should tell her/his offer (Turn 4). Jennifer can accept this offer or make a counteroffer (Turn 5). If Jennifer makes a counteroffer, she asks whether the human negotiator accepts this offer. The human negotiator should say "Yes" to accept this offer; otherwise, she/he should say "No" to continue negotiation with another offer (Turn 6). Afterward, the process continues in a turn-taking fashion (Turn 1–6) until having an agreement or reaching a predefined deadline.

The underlying protocol is flexible enough for enabling agents to perform different types of bilateral negotiation. While they may negotiate on their holiday (e.g., location, duration, activities), they can also negotiate how to allocate a set of resources between them. The proposed speech-based protocol consists of four fundamental phases as follows:

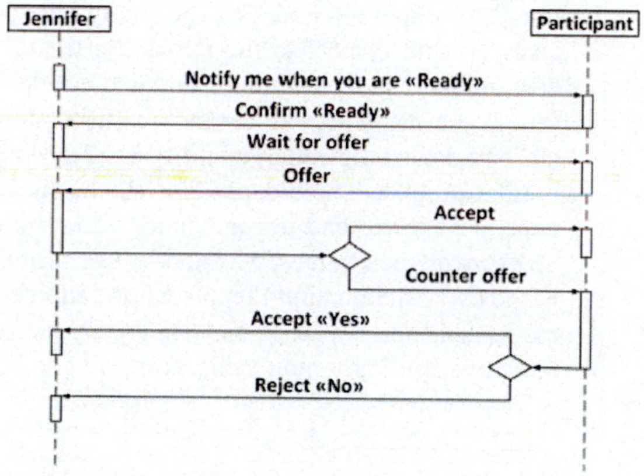

Fig. 1 FIPA representation of speech-based protocol

- **Notification Phase (Turn 1–3)**: Human negotiators may sometimes think out loud and what they said can be perceived as an offer by Jennifer. To avoid such misunderstanding, SHRNP ensures when exactly the human negotiator makes her/his offer by confirming when they are ready to make their offers so that Jennifer can process the right utterances to gather her opponent's offer.
- **Offering Phase for Human Agent (Turn 4)**: Human negotiator makes his/her offer.
- **Robot-Response Phase (Turn 5)**: Jennifer evaluates her opponent's offer and either accepts it or makes a counteroffer.
- **Human-Response Phase (Turn 6)**: If Jennifer makes a counteroffer, the human agent should inform whether she/he accepts/rejects the given offer.

Jennifer uses speech recognition to perceive what the other party says, and text-to-speech technology to speak to the human negotiator. For the fluidity of the conversation, our framework aims not to restrict the user with a set of predefined words. Therefore, Jennifer uses dictation instead of grammar-based speech recognition. That is, Jennifer listens to the human negotiator until she or he stops her/his speech. With the help of the speech recognition tool, Jennifer translates her opponent's speech into a set of words. Afterward, she processes the recognized words regarding the given phase. In notification and human-response phases, our agent tries to find a predefined keyword (e.g., "ready," "yes," or "no") in the given set and ignores other words. In the offering phase for human agent, the opponent may say her/his offer in a different way (e.g., use a different order). Therefore, our agent should process the given set of recognized words and convert them into a valid offer. In the following part, we introduce a negotiation strategy using a variant of time-dependent bidding tactic.

2.2 *Time-Dependent Stochastic Bidding Tactic*

Time-dependent concession strategies have been widely used in automated negotiation [22]. When a negotiating agent employs such a concession strategy, its behavior changes with respect to the remaining time. That is, the agent has a tendency to demand more at the beginning and to concede over time. The target utility at a given time is estimated according to a time-dependent function and a bid that has a utility close to the estimated target utility is offered by the agent.

The proposed time-dependent stochastic bidding tactic (TSBT) defines time-dependent lower and upper bounds and randomly generates a bid between them. To estimate their values, we adopt to use a time-dependent concession function proposed by [24]. Equation 1 represents the adopted concession function where t denotes the scaled time $t \in [0, 1]$ and P_0, P_1, P_2 are the maximum value, the curvature of the curve, and minimum value, respectively.[2]

[2]For the lower bound, P_0, P_1, and P_2 are 0.94, 0.5, and 0.4, respectively, and for the upper bound they are 1, 0.9, and 0.7, respectively, in our experiments as seen in Fig. 1.

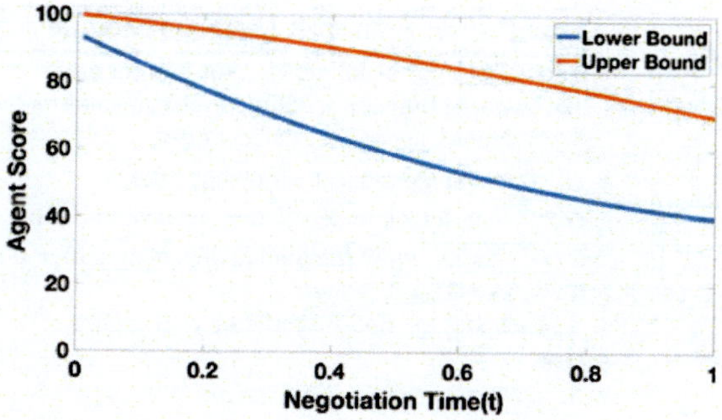

Fig. 2 Time-dependent lower and upper bounds

$$TU(t) = (1-t)^2 \times P_0 + 2 \times (1-t) \times t \times P_1 + t^2 \times P_2 \quad (1)$$

It is worth noting that the adaptive lower and upper bounds correspond to *Conceder* and *Boulware* behavior, respectively. Recall that Conceder agent concedes fast during the negotiation while Boulware agent hardly concedes until the deadline. As seen on Fig. 2, our agent may switch its strategy between these tactics stochastically. Consequently, the human opponent may not easily guess our agent's behavior.

2.3 Negotiation Strategy

Algorithm 1 shows how Jennifer makes her decisions during the negotiation. Jennifer checks whether the deadline is reached; if so, she ends the negotiation (Line 1–2). Otherwise, she generates her offer according to her bidding tactic (Line 3). If the utility of the opponent's current bid is higher than or equal to the utility of Jennifer's incoming offer, she accepts the given offer (Line 4–5). Otherwise, Jennifer makes her counteroffer (Line 6).

Afterward, Jennifer decides her attitude toward her opponent. Jennifer pushes her opponent as the deadline approaches (Line 9–11). There is a predefined time, T_Θ to warn the opponent once. If it is reached, Jennifer tells the opponent to hurry up as specified in Table 1. If the opponents make a humiliating offer, which is not acceptable at all, Jennifer feels "offended" (Line 12–13). That is when the utility of the given offer is less than the reservation utility – the minimum acceptable utility.

Otherwise, Jennifer calculates the utility change in her opponent's subsequent offers (Line 14) and decides her attitude considering the given offer and her opponent's move (e.g., concession, silent, selfish). As specified in Table 1, a mild behavior is adopted by Jennifer if Jennifer thinks that they are approaching a consensus. That happens when Jennifer employs TSBT and the utility of her opponent's offer for Jennifer, $U(O_h^{t_{cur}})$ is higher than or equal to the estimated lower threshold, LT.

Algorithm 1: Jennifer's Decision Module

Data: $T_{deadline}$: deadline, t_{cur}: the current time,
T_{Θ}: warning time for deadline, $tactic$: Jennifer's bidding tactic
$O_h^{t_{cur}}$: human opponent's current offer,
$O_h^{t_{prev}}$: human opponent's previous offer,
$O_j^{t_{cur}}$: Jennifer's counteroffer, R: reservation utility,
$U(O_h^t)$ the utility of the human opponent's offer at time t for Jennifer,

if $t_{cur} >= T_{deadline}$ **then**
  Behavior $\leftarrow$ End-Negotiation ;
else
  $O_j^{t_{cur}} \leftarrow$ generateBid($tactic$);
  if $U(O_j^{t_{cur}}) \leq U(O_h^{t_{cur}})$ **then**
    Behavior $\leftarrow$ Accept ;
  else
    Make $O_j^{t_{cur}}$;
    if $O_h^{t_{prev}} == null$ **then**
      isHurryUp $\leftarrow$ false ;
    else
      if *isHurryUp = false &* $T_{\Theta} <= t_{cur}$ **then**
        Behavior $\leftarrow$ Hurry-up ;
        isHurryUp $\leftarrow$ true ;
      else
        if $U(O_h^{t_{cur}}) < R$ **then**
          Behavior $\leftarrow$ Offended ;
        else
          $\Delta U \leftarrow U(O_h^{t_{cur}}) - U(O_h^{t_{prev}})$;
          Behavior $\leftarrow$ getMood($\Delta U, O_h^{t_{cur}}$) ;
        end
      end
    end
  end
end

When the opponent does not change the utility of its offer, it corresponds to a neutral behavior. If the opponent concedes ($\Delta U > 0$), Jennifer feels pleasant. On the other hand, if the opponent makes a selfish move (e.g., decreasing Jennifer's utility), she shows her dissatisfaction by saying that she did not like her opponent's offer. Table 1 indicates what Jennifer says to her human opponent in each case.

Table 1 Argument decision matrix

Case	Behavior	Arguments
$U(O_h^{t_{cur}}) < R$	Offended	It is not acceptable!
$\Delta U < 0$	Dissatisfied	I don't like your offer You should revise it
$\Delta U = 0$	Neutral	Himm
$\Delta U > 0$	Pleasant	It is getting better But not enough
$U(O_h^{t_{cur}}) >= LT$	Mild	I like your offer but you can increase a little bit
$U(O_h^{t_{cur}}) >= U(O_j^{t_{cur}})$	Acceptance	Yes, I accept your offer!
$T_\Theta <= t_{cur}$	Hurry up	Hurry up! We need to find an agreement soon
$t_{cur} >= T_{deadline}$	Time's up	Let's stop! We cannot reach an agreement

3 Evaluation

In order to evaluate the performance of the developed negotiation strategy with TSBT, we design a user experiment. Determining the basic design structure is a crucial task especially in HRN setting. In the following part, our experiment design and our findings are given.

3.1 Experimental Setup

We have recruited 30 participants (i.e., university students and faculty members; 19 Male, 11 Female; Median age: 23) for conducting our human-robot experiments. It is worth noting that the experiment was approved by Özyeğin University Research Ethics committee (REC) and participants that gave informed consent. Our main aim is to investigate whether the robot can perform at least as well as human negotiators in the given negotiation task. Therefore, we asked the volunteer participants to negotiate with Jennifer and then analyzed the negotiation outcomes elaborately.

In the experiment, a negotiation scenario is given to each participant and as a role-playing game, they are asked to study their preference profiles and the interaction protocol elaborately before their negotiation. Apart from the given negotiation scenario, an easy negotiation scenario has been created for the training session. The participants watched a video of a training negotiation session; afterward, they do a five-minute negotiation training session.

After the training session, the preference profile for the negotiation session is studied by the participant and then they negotiate with Jennifer for up to 10 min. The

Table 2 Preference profiles for negotiation sessions

Items	Jennifer's profile	Human's profile
Hammer	6	13
Container	22	20
Knife	5	10
Match	20	22
Compass	13	5
Medicine	7	6
Food	17	7
Rope	10	17

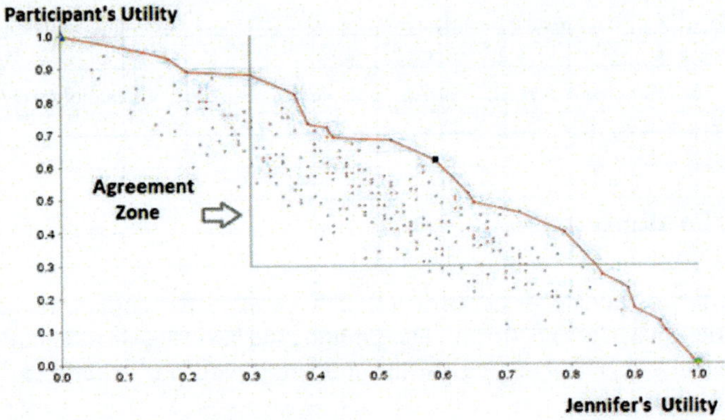

Fig. 3 Outcome space for negotiating parties

deadline for each negotiation is set as 10 min. If there is no agreement within 10 min, both parties receive zero points. Note that the aim of the participants is to receive at least 30 points out of 100. The participants are encouraged to maximize their score by pointing out that the participants with the highest score will win a gift card from a well-known coffee brand. Thus, the participants take their negotiation seriously.

According to our scenario, our participants need to negotiate with Jennifer on resource allocation in order to survive in a deserted island. There are eight indivisible items: Some of them will be given to the participant and the rest of them will be taken by Jennifer. Note that human participants ask for what they would like to get and Jennifer offers what items to be given to the participants in order to avoid misunderstanding. In other words, the negotiation is on what items would be given to the participant. Table 2 shows these items and their scores for Jennifer and her human counterpart. Figure 3 shows the utilities of each possible bid in the given scenario as well as the agreement zone. It can be seen that the bargaining power of two parties is almost the same.

Fig. 4 Experiment setup

It is worth noting that the participants only know their own scores and they are informed that Jennifer does not know their scores too.

As seen in Fig. 4, participants are allowed to use a paper to take their notes and their phones to check the remaining time. They keep current preference profile and interaction flowchart with them during their negotiation. Their negotiation session is recorded so as to check the quality of speech recognition with our detailed log files. At the end of their negotiation, each participant is asked to fill in a questionnaire form about their negotiation experience with Jennifer.

3.2 Experiment Results

Out of 30 negotiations, 26 negotiations ended up with an agreement while only 4 of them failed. In other words, participants reached an agreement in 86.7% of negotiations. Table 3 shows the detailed results of each successful negotiation in our experiment. First column shows negotiation session id and the following five columns indicate the percentages of human negotiator's attitude perceived by the robot as it is described in Table 1. For instance, "offended" indicates the percentage of offensive offers made by the human negotiator. The seventh and eighth columns show the score received by our agent (Jennifer) and the score gained by the human participant, respectively. The last column shows the normalized agreement time [0, 1].

Jennifer beats the human participants in approximately 58% of the negotiations (15 out of 26). Furthermore, we test the following null hypothesis, H_0: There is no difference between the gained score by human negotiator and the score collected by the robot. We applied t-tailed test for two paired samples on agent score and user score. Since it is observed that $t = 2.937 > tc = 1.708$ and $p = 0.0035 < 0.05$, it is concluded that the null hypothesis is rejected. In other words, the scores for the agent and human participants are significantly statistically different at 95% confidence interval. That is, we can conclude that Jennifer outperformed human negotiators on average (60.96 vs. 49.42) as also seen in Fig. 5.

Table 3 Analysis of successful negotiations

ID	Offended	Dissatisfied	Neutral	Pleasant	Mild	Agent score	User score	Agreement time
1	0.0	7.5	0.0	37.5	25.0	79	40	0.757
2	0.0	50.0	0.0	50.0	0.0	77	36	0.505
3	0.0	66.7	0.0	33.3	0.0	75	32	0.272
4	0.0	0.0	0.0	0.0	100.0	75	32	0.473
5	0.0	12.5	0.0	37.5	50.0	72	30	0.846
6	0.0	100.0	0.0	0.0	0.0	69	34	0.492
7	0.0	25.0	0.0	50.0	25.0	68	37	0.466
8	0.0	40.0	40.0	20.0	0.0	65	49	0.478
9	0.0	20.0	0.0	40.0	40.0	65	49	0.644
10	0.0	0.0	0.0	66.7	33.3	65	49	0.743
11	0.0	20.0	20.0	40.0	20.0	63	47	0.545
12	37.5	12.5	12.5	25.0	12.5	62	50	0.688
13	0.0	44.4	0.0	44.4	11.1	60	49	0.618
14	14.3	28.6	14.3	42.9	0.0	59	62	0.693
15	0.0	60.0	0.0	20.0	20.0	59	62	0.553
16	66.7	0.0	0.0	22.2	11.1	59	62	0.702
17	0.0	50.0	0.0	50.0	0.0	57	58	0.556
18	0.0	44.4	33.3	22.2	0.0	57	44	0.647
19	0.0	11.1	22.2	44.4	22.2	57	58	0.749
20	0.0	45.5	9.1	27.3	18.2	55	43	0.724
21	0.0	75.0	0.0	25.0	0.0	50	66	0.755
22	0.0	62.5	0.0	37.5	0.0	48	59	0.807
23	0.0	71.4	0.0	28.6	0.0	48	59	0.845
24	9.1	54.5	0.0	18.2	18.2	47	59	0.982
25	76.9	0.0	0.0	15.4	7.7	47	65	0.924
26	23.1	23.1	0.0	53.8	0.0	47	54	0.838
Mean	8.75	36.72	5.82	32.77	15.94	60.96	49.42	0.665

As far as the agreement time is concerned, it is seen that it took about 6–7 min to find an agreement with Jennifer. Due to the nature of our agent's bidding strategy, the human participants who were patient to wait longer gained higher score. Recall that Jennifer concedes stochastically over time.

When the behavior of best performing human negotiator is studied, it is observed that 75% of his/her moves is dissatisfying ($\Delta U < 0$). On the contrary, the worst performing human negotiator made mostly pleasant and mild behavior (37.5% and 50%, respectively).

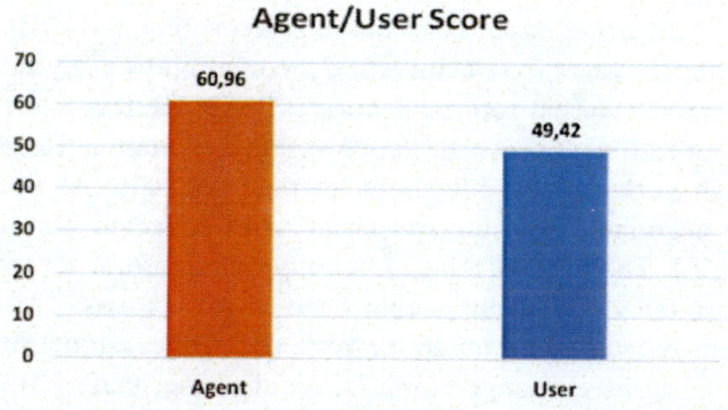

Fig. 5 Average scores in agreements

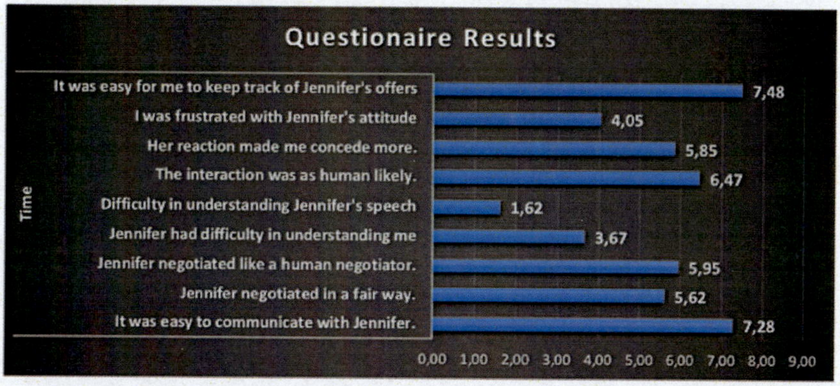

Fig. 6 Questionnaire

Figure 6 shows the average ratings given by the users to our questionnaire consisting of 9-point scaled questions after their negotiation (1 for strongly disagreement whereas 9 for strongly agreement). The resulting average ratings of the more positively structured statements such as "Her gestures were mostly consistent with the situation." were satisfyingly high. Besides, some of the negatively structured statements such as "I was frustrated with Jennifer's attitude." had reasonably low scores (i.e., disagree) as positive feedback.

4 Related Work

In this section, we review most recent works on human-agent negotiation. Designing agents that are able to negotiate with humans require considering some human factors and dynamics [13]. For example, the fairness of the offers might have a significant influence on human negotiator's decision making.

Most of the agent-based negotiation systems (e.g.,[14–16]) use text-based communication. However, communication medium plays a key role in human negotiations and mostly verbal communication is preferred. It is more natural and effective to communicate via speech in human-agent negotiation. Recently, DeVault, Mell, and Gratch worked on establishing a fluent conversation between a virtual agent and a human negotiator by using speech libraries collected from human-human negotiations [25]. The agent itself is not fully autonomous; the speech and high-level behavior of the virtual agent is controlled by two experts.

In agent-based negotiation framework, during the negotiation in addition to offers, arguments can be exchanged to persuade the other party [26, 27]. In recent years, negotiation frameworks support to exchange arguments [14, 15]. IAGO has been developed for human-agent negotiations in which parties can exchange offers, arguments, and emotional expressions. Note that the agent uses a predefined set of utterances during its negotiation. Moreover, Mell et al. investigate whether using the arguments indicating the appreciation to the opponent has an impact on the negotiation. [28]. In another study, by applying neural networks and reinforcement learning on the dialogues collected from human-human negotiations, it is aimed to learn how to use dialogues effectively in negotiation [29].

Although there are a variety of works on human-virtual agent negotiation, there are relatively less work on negotiating robots. Bevan and Fraser studied experimentally whether or not handshaking before the negotiation has a comprising effect in negotiation. [30] examined experimentally whether or not the use of guilty expression by a robot has an effect on its opponent's compromise. In almost all of those works [30], robots are remotely controlled by a human. To best of our knowledge, there is an urgent need to develop fully autonomous negotiating humanoid robots.

5 Conclusion

This work introduces a novel negotiation scheme in which a humanoid robot negotiates effectively with a human counterpart via speech. Our experimental results showed that our robot can negotiate at least as well as human counterpart on average. Even Jennifer managed to outperform the human participants although it employs a time-based concession strategy.

As a future work, we are planning to develop more sophisticated negotiation strategies and compare their performance with the performance of the time-based concession strategy. Moreover, we would like to investigate the effect of other factors such as "gesture" on the negotiation outcome. Furthermore, it would be interesting to study the cultural differences in human-robot negotiation as [31] do.

References

1. M. Johansson, T. Hori, G. Skantze, A. Höthker, J. Gustafson, Making turn-taking decisions for an active listening robot for memory training, in*ICSR*, volume 9979 of *Lecture Notes in Computer Science* (2016), pp. 940–949
2. J. Xu, J. Broekens, K.H. Hindriks, M. Neerincx, Mood expression of a robotic teacher on students, in *Proceedings of the 2014 IEEE/RSJ International Conference on Intelligent Robots and Systems (IROS)* (2014), pp. 2614–2620
3. J. Mell, J. Gratch, T. Baarslag, R. Aydoğan, C.M. Jonker, Results of the first annual human-agent league of the automated negotiating agents competition, in *Proceedings of the 18th International Conference on Intelligent Virtual Agents, IVA '18* (ACM, New York, NY, USA, 2018), pp. 23–28
4. B. An, N. Gatti, V.R. Lesser, Extending alternating-offers bargaining in one-to-many and many-to-many settings, in *Proceedings of the 2009 IEEE/WIC/ACM International Conference on Intelligent Agent Technology*, vol. 2 (2009), pp. 423–426
5. R. Aydogan, T. Baarslag, K.V. Hindriks, C.M. Jonker, P. Yolum, Heuristics for using CP-nets in utility-based negotiation without knowing utilities. Knowl. Inf. Syst. **45**(2), 357–388 (2015)
6. R. Aydoğan, D. Festen, K.V. Hindriks, C.M. Jonker, Alternating offers protocols for multilateral negotiation, in K. Fujita, Q. Bai, T. Ito, M. Zhang, F. Ren, R. Aydoğan, R. Hadfi (eds.), *Modern Approaches to Agent-based Complex Automated Negotiation* (Springer, 2017), pp. 153–167
7. S. Fatima, S. Kraus, M. Wooldridge, *Principles of Automated Negotiation* (Cambridge University Press, 2014)
8. S.S Fatima, M. Wooldridge, N.R. Jennings, An agenda-based framework for multi-issue negotiation. Artif. Intel. **152**(1), 1–45 (2004)
9. G.E. Kersten, G. Lo, Aspire: an integrated negotiation support system and software agents for e-business negotiation. Int. J. Internet Enterp. Manage. **1**(3), 293–315 (2003)
10. R. Lin, S. Kraus, T. Baarslag, D. Tykhonov, K. Hindriks, Catholijn M. Jonker, Genius: an integrated environment for supporting the design of generic automated negotiators. Comput. Intell. **30**(1), 48–70 (2014)
11. I. Marsa-Maestre, M. Klein, C.M. Jonker, R. Aydoğan, From problems to protocols: towards a negotiation handbook. Decision Support Syst. **60**, 39–54 (2014) (Automated Negotiation Technologies and their Applications)
12. V. Sanchez-Anguix, R. Aydogan, V. Julian, C. Jonker, Unanimously acceptable agreements for negotiation teams in unpredictable domains. Electron. Comm. Res. Appl. **13**(4), 243–265 (2014)
13. R. Lin, Sarit Kraus, Can automated agents proficiently negotiate with humans? Commun. ACM **53**(1), 78–88 (2010)
14. A. Rosenfeld, I. Zuckerman, E. Segal-Halevi, O. Drein, S. Kraus, Negochat: a chat-based negotiation agent, in *AAMAS* (IFAAMAS/ACM, 2014), pp. 525–532
15. J. Mell, J. Gratch, IAGO: interactive arbitration guide online (demonstration), in *AAMAS* (2016), pp. 1510–1512
16. C.M. Jonker, R. Aydo Yan, T. Baarslag, J. Broekens, C.A. Detweiler, K.V. Hindriks, A. Huldtgren, W. Pasman, An introduction to the pocket negotiator: a general purpose negotiation support system, in *EUMAS 2016* (Springer, 2016), pp. 13–27
17. A.L. Thomaz, C. Chao, Turn-taking based on information flow for fluent human-robot interaction. AI Magazine **32**(4), 53–63 (2011)
18. C. Bevan, D. Stanton Fraser, Shaking hands and cooperation in tele-present human-robot negotiation, in *HRI* (2015), pp. 247–254
19. T. Baarslag, E.H. Gerding, R. Aydogan, M.C. Schraefel, Optimal negotiation decision functions in time-sensitive domains, in *IEEE/WIC/ACM WI-IAT* (2015), pp. 190–197
20. S. Kawaguchi, K. Fujita, T. Ito, Agentk: compromising strategy based on estimated maximum utility for automated negotiating agents, in *New Trends in Agent-Based Complex Automated Negotiations* (Springer, 2012), pp. 137–144

21. C.R. Williams, V. Robu, E.H. Gerding, N.R. Jennings, Iamhaggler2011: a Gaussian process regression based negotiation agent, in *Complex Automated Negotiations: Theories, Models, and Software Competitions* (Springer, 2013), pp. 209–212
22. P. Faratin, C. Sierra, N.R. Jennings, Negotiation decision functions for autonomous agents. Robot. Auton. Syst. **24**(3–4), 159–182 (1998)
23. Ariel Rubinstein, Perfect equilibrium in a bargaining model. Econometrica **50**(1), 97–109 (1982)
24. R. Vahidov, G.E. Kersten, B. Yu, Human-agent negotiations: the impact agents' concession schedule and task complexity on agreements, in *Proceedings of the 50th Hawaii International Conference on System Sciences (HICSS)* (2017), pp. 412–420
25. D. DeVault, J. Mell, J. Gratch, Toward natural turn-taking in a virtual human negotiation agent, in *AAAI Spring Symposium on Turn-taking and Coordination in Human-Machine Interaction* (AAAI Press, Palo Alto, California, March 2015), pp. 2–9
26. P. Pasquier, R. Hollands, I. Rahwan, F. Dignum, Liz Sonenberg, An empirical study of interest-based negotiation. Auton. Agents Multi-Agent Syst. **22**(2), 249–288 (2011)
27. I. Rahwan, S.D. Ramchurn, N.R. Jennings, P. McBurney, S. Parsons, Liz Sonenberg, Argum. Negot. Knowl. Eng. Rev. **18**(4), 343–375 (2003)
28. J. Mell, G.M. Lucas, J. Gratch, An effective conversation tactic for creating value over repeated negotiations, in *Proceedings of the 2015 International Conference on Autonomous Agents and Multiagent Systems* (AAMAS, Istanbul, Turkey, 4–8 May 2015), pp. 1567–1576
29. M. Lewis, D. Yarats, Y. Dauphin, D. Parikh, D. Batra, Deal or no deal? end-to-end learning of negotiation dialogues, in *EMNLP* (Association for Computational Linguistics, 2017), pp. 2443–2453
30. B. Stoll, C. Edwards, A. Edwards, Why aren't you a sassy little thing: the effects of robot-enacted guilt trips on credibility and consensus in a negotiation. Commun. Stud. **67**(5), 530–547 (2016)
31. G. Haim, Y. (Kobi) Gal, M. Gelfand, S. Kraus, A cultural sensitive agent for human-computer negotiation, in *AAMAS* (2012), pp. 451–458

Effect of Morality for Automated Negotiating Agents: A Preliminary Result

Takayuki Ito

Abstract In this paper, we analyze the effect of morality in agent-based automated negotiations. Much attention has been given to the axiomatic decision-making model based on self-control under the presence of temptation. Here, self-control and shame are modeled as the mental costs of morality, and they are explicitly incorporated in the utility function. This is because the recent advances in behavioral economics have experimentally proved that real decision making is not simply based on a single value criterion but on multiple value criteria. Gul and Pesendorfer [8] first proposed a decision-making model based on self-control and temptation (thus called the GP model), and since then it has been extended. Dillenberger and Sadowski [6] proposed a shame (morality)-based utility model based on the GP model, in which an agent faces the trade-off between its selfishness and morality. In this paper, we apply a morality-based utility model to automated negotiation agents and analyze the effect of morality on automated negotiation and the related consensus-making.

1 Introduction

This paper discusses the effect of morality on automated negotiation between agents. Much attention has been given to the relation between rationality and morality in AI-related fields [4, 24]. When we assume that an agent acts in our real society, then it should make decisions that conform to the realities of people. Consequently, we need to incorporate morality in the agent's decision-making mechanism. However, although there has been much discussion on how to incorporate morality in a preference model, so far no concrete proposal has been put forward.

Automated negotiation has received much attention in the field of multi-agent systems [14, 17, 26, 28]. There is a rich literature on it, and automated negotiation

T. Ito (✉)
Nagoya Institute of Technology, Gokiso, Showa-ku, Nagoya 466-8555, Japan
e-mail: ito.takayuki@nitech.ac.jp
URL: http://www.itolab.nitech.ac.jp/ ito/

T. Ito et al. (eds.), *Advances in Automated Negotiations*, Studies in Computational Intelligence 905, https://doi.org/10.1007/978-981-15-5869-6_2

is one of the key concepts in multi-agent systems because negotiation is a central activity in human society.

In the field of decision making, much attention has been paid to the axiomatic utility models as a way to represent self-control under the pressure of temptation [6, 8, 18]. These approaches explicitly incorporate the mental cost for self-control or shame in a utility function. In particular, one proposed model [6] focused on the mental cost incurred when people face an individual decision with a consequence that could violate morality, which means a decision that brings shame. This is because it has been empirically proved that human decision making is done based not on a single criterion, as in classic economics, but on several value criteria, as taken up in experimental psychology and behavioral economics. In this paper, we employ a utility model with morality and shame in the automated negotiation framework and demonstrate its effect.

In classic economics and game theory, the most standard utility model is that of Von Neumann and Morgenstern (NM) [20]. They introduced six axiomatic demands that should be satisfied for a rational human's utility: completeness, transitivity, substitutability, decomposability, monotonicity, and continuity. They proved that there are such utility functions (called NM utility functions) that satisfy these six axioms. In using the NM utility model, an agent is assumed to be the expected utility maximizer. In general, many discussions are based on a single value criterion.

Furthermore, there have been many works on multi-criteria/attribute utility functions [15, 29] and time-dependent dynamic utility functions [27]. In the AI and MAS fields, a lot of work has been carried out on multi-attribute and multi-issue negotiation models [7, 12, 22]. However, none of these works considered a player's psychological conflict due to having multiple criteria.

Therefore, Gul and Pesendorfer [8] first proposed a decision-making model based on self-control and temptation (hereafter, the GP model). For example, let us assume that an agent, Alex, is on a diet and needs to make a decision over whether to have a hamburger or a salad for lunch. Because he loves hamburgers, he is tempted to go to a restaurant that has hamburgers on the menu. However, because he is on a diet, he chooses (make a decision) to go to another restaurant, which has salad but not hamburgers on its menu. This kind of decision making is called self-control under temptation.

The GP model succeeded in representing self-control under temptation. Dillenberger and Sadowski [6] proposed an extended model that handles "shame" when people make an individual decision under consideration of morality (i.e., a norm). Namely, a human is tempted to make a selfish decision while controlling her/himself based on a moral or normative rule. Dillenberger and Sadowski [6] identified "shame as the moral cost an individual experiences if instead of choosing an alternative that she perceives to be in accordance with a social norm (which might include, but is not limited to, considerations of fairness and altruism), she is observed choosing an alternative that favors her own material payoffs."

In their paper [6], they provided a concrete example of a utility function that employs the Nash solution in cooperative games as the moral criterion. Here, for agents who have this utility function, achieving an agreement itself is a moral act, as in the generalized view of Asian culture: "Harmony is to be valued." Thus, we are able to give agents a moral rule that says we should pursue a desirable agreement, such as the Nash solution, through agents.

This paper shows that, by introducing in the utility function a mental cost of violating morality, we can make a negotiation more likely to lead to agreement among agents. In other words, the moral standard of "Harmony is to be valued" would make agents more prone to agreement in automated negotiation.

In the literature of automated negotiation research [1, 12], there has been almost no comprehensive explanation of why agents compromise during negotiation. Compromising means decreasing one's own utility, which could run counter to the utility maximizer assumption. One standard explanation is that agents **need** to compromise because they will get zero utility if they cannot reach an agreement by the deadline. This explanation sounds rational. However, in this case, the reason the agent decreases its own utility for compromising is **an extrinsic motivation**. Agents need to decrease their own utilities even when they are utility maximizers. Time-dependent discounted utility might be another explanation, but how and how much an agent should discount its utility from time to time is a very difficult and open problem.

In this paper, we analyze the effect of a utility model that can incorporate the mental cost of shame in automated negotiation among agents. First, we introduce the Gul and Pesendorfer model (GP model), and then, we show Dillenberger's morality-based model that incorporates a mental cost for shame. Next, we analyze the effect of employing the morality-based utility model in bilateral automated negotiations. Finally, we show how this model differs from related work and then present our conclusions and future work.

2 GP Model: Self-control Under Temptation

The utility model based on self-control under temptation (GP model) was proposed by Gul and Pesendorfer [8]. This model is basically a two-period expected utility model. In period 1, an agent chooses a menu from a set of menus. Then, in period 2, it needs to make a selection from the menu chosen in period 1.

Let X be a compact metric space and $\delta(X)$ be the set of all measures on the Borel γ-algebra of X. We endow $\delta(X)$ with the topology of weak convergence and let Z be the set of non-empty, compact subsets of $\delta(X)$. We endow Z with the topology generated by the Hausdorff metric. Define $\alpha x + (1-\alpha)y := \{p = \alpha q + (1-\alpha)r : q \in x, r \in y\} fore, y \in Z, \alpha \in [0, 1]$. The set X represents the possible second-period alternatives, $\delta(X)$ represents the lotteries, and Z represents the object of choice in the first period.

The following axioms are standard assumptions in decision theories.

Axiom 1 **(Preference Relation)**: $\succeq$ is a complete and transitive binary relation.

Completeness means that, given two alternatives, the decision maker is required to prefer one to the other or to favor both ones equally. Namely, $A \succ B \vee B \succ A \vee A \sim B$ if there are alternatives A and B.

Transitivity means that if a decision maker prefers A to B and B to C, then he/she is required to prefer A to C when there are three alternatives. Namely, $(A \succ B) \wedge (B \succ C) \Rightarrow (A \succ C)$ when there are alternatives A, B, and C.

Axiom 2 (**Continuity**): $A, B, C \in Z$, if $A \succ B \succ C$, then there exists $\alpha, \beta \in (0, 1)$.

$$\alpha A + (1 - \alpha)C \succ B \succ \beta A + (1 - \beta)C$$

Axiom 3 (**Independency**): For $A, B, C \in Z$; $\alpha \in [0, 1]$,

$$A \succeq B \Leftrightarrow \alpha A + (1 - \alpha)C \succeq \alpha B + (1 - \alpha)C$$

The following axiom is the characteristic axiom for the model of self-control under temptation [8].

Axiom 4 (**Set-Betweeness**): For $A, B \in Z$,

$$A \succeq B \Rightarrow A \succeq A \cup B \succeq B$$

Under the condition of self-control under temptation, $A \succeq A \cup B$ is a feasible preference. In this case, taking the example presented in the introduction, A is a restaurant that only has salads on its menu, while B is a restaurant that serves hamburgers. Based on this condition, a decision maker chooses A using her/his self-control under the temptation to choose B.

Theorem 1 *[8] When a preference $\succeq$ is represented by any $A \in Z$ and continuous linear functions U, u, v,*

$$U(A) := \max_{x \in A}\{u(x) + v(x)\} - \max_{y \in A} v(y) \tag{1}$$

the preference $\succeq$ satisfies the above axioms 1 to 4.

Equation (1) is called a self-control utility [18]. The self-control utility U is characterized by two utility functions u and v. u (called a "commitment utility") implies a commitment to an opportunity in the opportunity sets of the first stage. v (called a "temptation utility") implies temptation to an alternative in the second stage. Here, when $u(x) > u(y)$ and $v(y) > v(x)$, then $U(\{x\}) > U(\{x, y\})$. Namely, a decision maker with this preference is tempted to choose y but commits her/himself to $\{x\}$.

If commitment is impossible, the decision maker selects a menu $\{x, y\}$. Then, in the second stage, she/he chooses y (by temptation) or x (by commitment). In this case, if $u(y) + v(y) > u(x) + v(x)$, she/he chooses y by temptation, while if $u(y) + v(y) < u(x) + v(x)$, she/he chooses x by commitment (self-control).

Here, the following Eq. (2) shows the cost for exerting self-control.

$$v(x) - \max_{\{x,y\}} v(x) = v(x) - v(y). \quad (2)$$

3 Utility Model with Shame and Social Norm

Dillenberger and Sadowski [6] extended the GP model to model a preference based on shame and social norms between two players. Here, the concept of shame is the mental cost that a decision maker feels in choosing an alternative according to her/his own selfish preference, which corresponds to the concept of temptation in the GP model.

The following examples from the literature show there is a mental cost to making a decision that defies morals, which is called shame. For example, in the well-know "dictator game," two players, A and B, divide up a certain amount of money. However, only one player, say A, can make a decision on how to divide the total money, say \$10. Namely, A can decide the amounts that both A and B will obtain.

In the dictator game, it is rational for A to claim the total amount of \$10 if she/he follows the utility maximization principle. However, in most of the social psychology experiments, it has been observed that the person who plays the role of A gives around \$0–\$5 to B [3]. This result is considered proof of the existence of altruism and fairness in human preferences.

However, Dana [5] further investigated a variant of the dictator game, where A can have the option to exit the game before B realizes the fact that they are playing the game itself. They found that one-third of the participants gave \$9 to A (themselves) and \$0 to B. Let us represent this allocation as (\$9, \$0). If we believe the altruism rationale, then it must be that (\$9, \$1) is preferred to (\$9, \$0). On the other hand, if we believe in pure utility maximizing, (\$10, \$0) must be preferred.

From the above observation, Dillenberger and Sadowski [6] proposed an interpretation that a human controls her/his selfishness when she/he can behave **socially**. Then, they defined **shame** as "the moral cost an individual experiences if instead of choosing an alternative that she perceives to be in accordance with a social norm (which might include, but is not limited to, considerations of fairness and altruism), she is observed choosing an alternative that favors her own material payoffs."

The following axioms and definitions are based on Dillenberger and Sadowski [6]. Here, $a = (a_1, a_2)$ means that when alternative a is selected, then player 1 obtains payoff a_1 and player 2 obtains payoff a_2.

Axiom 5 (**Weak order**): $\succeq$ is complete and transitive.

Axiom 6 (**Continuity**): $\succeq$ is continuous.

Axiom 7 (**Strong left betweenness**): If $A \succeq B$, then $A \succeq A \cup B$. Furthermore, if $A \succ B$ and there exists C such that $A \cup C \succ A \cup B \cup C$, then $A \succ A \cup B$.

Definition 2 We say that the decision maker is **susceptible** to shame if there exist A and B such that $A \succ A \cup B$.

This definition implies the same situation where the decision maker is tempted to choose B under the self-control utility model.

Definition 3 If the decision maker is susceptible to shame, then we say that the decision maker deems b to be normatively better than a, written $b \succ_n a$, if $\exists A \in Z$; $a \in A$, such that $A \succ A \cup \{b\}$.

This definition shows the situation where the decision maker, if she/he prefers an opportunity set that does not include b to a set that does include b, normatively prefers b to a. This is the same as the commitment and self-control models.

Axiom 8 (**Ranking**): $\succ_n$ is an asymmetric and negatively transitive binary relation.

Axiom 8 means that we do not assume that a contributes to shame in A or b contributes to shame in B when there are two alternatives a and b and two menus A and B, where $\{a, b\} \subset A \cap B$.

Axiom 9 (**Pareto**): If $b \geq a$ and $b \neq a$, then $b \succ_n a$.

This axiom says if both players can obtain better payoffs by the alternative, then it is also normatively better.

Axiom 10 (**Compensation**): For all a, b, there exist x, y such that both $(a_1, x) \succ_n (b_1, b_2)$ and $(y, a_2) \succ_n (b_1, b_2)$.

Axiom 11 (**Selfishness**): If $a_1 > b_1$ and $a \succ_n b$, then $\{a\} \succ \{b\}$.

Definition 4 Let us assume that f and h are functions on X^2. We say that h is more selfish than f if for all a and for all δ_1 and δ_2, such that $(a_1 - \delta_1, a_2 - \delta_2) \in X^2$, where

(i) $h(a) = h(a_1 - \delta_1, a_2 + \delta_2)$ implies $f(a) \leq f(a_1 - \delta_1, a_2 + \delta_2)$;

(ii) $h(a) = h(a_1 + \delta_1, a_2 - \delta_2)$ implies $f(a) \geq f(a_1 + \delta_1, a_2 - \delta_2)$ with strict inequality for at least one pair δ_1, δ_2.

This definition means that the slope of a level curve of h in the (a_1, a_2) plane is at any point weakly greater than that of f.

Definition 5 A function $\varphi : X^2 \to \mathbb{R}$ is called a $subjectivenormfunction$ if it is strictly increasing and satisfies $\sup_{x \in X} \varphi(x, y) > \varphi(b)$ and $\sup_{x \in X} \varphi(y, x) > \varphi(b)$ for all $y \in X$ and $b \in X^2$.

Theorem 6 (**DS model** *[6]*): *The relations $\succeq$ and $\succ_n$ satisfy Axioms 5–11, respectively, if and only if there exist (i) a continuous subjective norm function φ, (ii) a continuous function $u : X^2 \to \mathbb{R}$, which is weakly increasing and more selfish than φ, and (iii) a continuous function $g : X^2 \times \varphi(X^2) \to \mathbb{R}$, which is strictly increasing* *in its second argument and satisfies* $g(a, x) = 0$ *whenever* $\varphi(a) = x$*, such that the function* $U : K \to \mathbb{R}$*, defined as*

$$U(A) = \max_{a \in A}[u(a) - g(a, max_{b \in A}\varphi(b))]. \tag{3}$$

Here, u and φ are NM utility functions. $u(a)$ is the selfish utility, while $g(a, \max_{b \in A} \varphi(b))$ means the mental cost (shame) of violating the social norm.

4 Effect in Negotiation Among Agents

4.1 A Model as Resource Allocation

When using a utility that includes shame as a social norm and a moral stance, there could be cases in which agents are able to make more agreements than using a standard utility model like the NM model.

In this section, we consider a resource allocation problem between two agents [2, 11, 25] as a case where a moral-based utility works effectively. In such a resource allocation problem between agents, we need to solve the allocation of multiple different items between the two agents, who have different utility functions. Each agent does not open its own utility functions to the other. In particular, the objective is to achieve a desirable allocation, such as a Pareto-efficient allocation. In this paper, we analyze the effect of a morality-based utility by using a concrete example 1 of the resource allocation problem [11, 25].

The resource allocation problem can be defined by a set of agents $N = \{1, 2\}$, a set of resources $R = \{r_1, r_2, \ldots, r_k\}, k \geq 2$, allocation $(a_1, a_2), a_1 \wedge a_2 = \emptyset, a_1 \in R, a_2 \in R$, and agent i's payoff function $p_i : a_i \to \mathbb{R}$. Based on this definition, a previous work [2] proposed a simple alternating-offer-based multi-stage negotiation protocol. Moreover, negotiation tree-based multi-stage negotiation protocols have also been proposed [11, 25].

In this paper, we consider a two-stage negotiation protocol, but it would be possible to extend this to an n-stage negotiation protocol. Agents try to make an agreement on opportunity sets in the first stage, and then, they try to make an agreement on the concrete allocation of the goods in the second stage. In this paper, we focus on the first-stage agreement over opportunity sets. For the second-stage agreement, more concrete protocols [2, 11, 25] can be directly employed.

Let us consider the following two examples of opportunity sets. Table 2 shows the opportunity set A, which is biased to each agent. Table 3 shows the opportunity set B, which is unbiased and nearly fair. Here, $A \cup B$ is equivalent to the situation of Table 1.

Table 1 Example of resource allocation

Allocation	1's payoff	2's payoff
({}, {*abcd*})	0	20
({*a*}, {*bcd*})	6	12
({*b*}, {*acd*})	8	14
({*c*}, {*abd*})	5	16
({*d*}, {*abc*})	7	18
({*ab*}, {*cd*})	9	8
({*ac*}, {*bd*})	10	9
({*ad*}, {*bc*})	11	8
({*bc*}, {*ad*})	10	8
({*bd*}, {*ac*})	11	13
({*cd*}, {*ab*})	12	15
({*abc*}, {*d*})	12	6
({*abd*}, {*c*})	14	7
({*acd*}, {*b*})	16	7
({*bcd*}, {*a*})	14	8
({*abcd*}, {})	20	0

Table 2 Opportunity set A

Allocation	1's payoff	2's payoff
({}, {*abcd*})	0	20
({*a*}, {*bcd*})	6	12
({*b*}, {*acd*})	8	14
({*c*}, {*abd*})	5	16
({*d*}, {*abc*})	7	18
({*abc*}, {*d*})	12	6
({*abd*}, {*c*})	14	7
({*acd*}, {*b*})	16	7
({*bcd*}, {*a*})	14	8
({*abcd*}, {})	20	0

Table 3 Opportunity set B

Allocation	1's payoff	2's payoff
({*ab*}, {*cd*})	9	8
({*ac*}, {*bd*})	10	9
({*ad*}, {*bc*})	11	8
({*bc*}, {*ad*})	10	8
({*bd*}, {*ac*})	11	13
({*cd*}, {*ab*})	12	15

4.2 The Case of a Selfish Utility Function

First, let us consider selfish agents. We define agent i's utility function as shown in equation (4):

$$U_i(A) = \max_{a \in A} u_i(a). \tag{4}$$

Obviously, there could be many other ways to define a selfish utility function. The above definition is one of the simpler definitions.

For example, let us define $u_i(a) = 2a_i$, where 2 could be any other fixed number). When we compare the opportunity sets A and B, then $U_1(A) = 20$, $U_1(B) = 12$, $U_2(A) = 20$ and $U_2(B) = 15$. Therefore, as the first-stage decision, it is rational for each agent to select the "biased" opportunity set A.

However, when we consider the second-stage decision, the following behavior will occur. Agent 1 can obtain $U_1(A) = 20$, and its concrete allocation will be $(\{a, b, c, d\}, \{\})$, where agent 1 gets everything and agent 2 gets nothing. On the other hand, agent 2 can get $U_2(A) = 20$, and its allocation will be $(\{\}, \{a, b, c, d\})$, where agent 1 gets nothing and agent 2 gets everything. Namely, because they choose opportunity set A, which is biased for each agent, they cannot reach an agreement in the second stage.

In the existing work on negotiation (e.g., [1]), if agents cannot reach an agreement, they are **forced** to concede and decrease their own utility, even if they are assumed to be the utility maximizer. How to design concession strategies remains an interesting research problem [1].

However, there has been little serious discussion on why agents decrease their own utility for conceding. **The negotiation protocol** pushes agents to concede and get a better agreement as a group. In this paper, we propose a morality-based utility model for automated negotiation, where agents can reach an agreement with a morality-based utility function that can satisfy utility maximization as well as the several axioms shown above.

4.3 The Case of a Utility Function with Morality

In this section, we consider a negotiation model based on a morality-based utility function. Let us define agent i's utility function as follows:

$$U_i(A) = \max_{a \in A}[u_i(a) - g_i(a, \max_{b \in A} \varphi(b))].$$

Let us define $u_i(a) = 2a_i$, $\varphi_i(a) = (a_1 + 1)(a_2 + 1)$, and $g_i(a, y) = y - \varphi(a)$ (from a previous example [6]).

Actually, $\max_{b \in A} \varphi(b) = (b_1 + 1)(b_2 + 1)$ represents the Nash solution [19], which is the desirable bargaining solution that can be considered a moral or social norm (+1 in each bracket is simply a fixed number and can be omitted). Here, we can employ any other formula that can be considered a moral or social norm.

Agent 1's utility for the opportunity set A is $U_1(A) = 16$, calculated as follows:

$$\begin{aligned} U_i(A) &= \max_{a \in A}[2a_1 - (\max_{b \in A}(b_1 + 1)(b_2 + 1) - (a_1 + 1)(a_2 + 1))], \\ &= \max_{a \in A}[2a_1 - (152 - (a_1 + 1)(a_2 + 1))], \\ &= \max_{a \in A}[2a_1 + (a_1 + 1)(a_2 + 1)] - 152, \\ &= 16. \end{aligned}$$

Here, in $\max_{b \in A}(b_1 + 1)(b_2 + 1)$, the allocation $(\{d\}, \{a, b, c\})$ maximizes $(b_1 + 1)(b_2 + 1)$. Thus, for agent 1, the allocation that maximizes the moral or social norm, including agent 2's allocation, is $(\{d\}, \{a, b, c\})$, and its utility becomes 152. In addition, $\max_{a \in A}[2a_1 + (a_1 + 1)(a_2 + 1)]$ consists of the selfish utility that maximizes the allocation and morality-based utility, and it is calculated as 168.

In the same way, the following utilities can be calculated. $U_1(B) = 24$, and its allocation is $(\{c, d\}, \{a, b\})$. $U_2(A) = 30$, and its allocation is $(\{d\}, \{a, b, c\})$. $U_2(B) = 30$, and its allocation is $(\{c, d\}, \{a, b\})$.

Namely, for agent 1, $U_1(A) = 16 < 24 = U_1(B)$, while for agent 2 $U_2(A) = 30 = 30 = U_2(B)$. For these two agents as a group, the opportunity set B is better. In particular, in this example, these agents can reach an agreement on $(\{c, d\}, \{a, b\})$ at the second stage.

As we described with a concrete example, if agents have morality-based utility functions, they can reach an agreement. But, for the same example, agents that have selfish utility functions cannot reach an agreement. Without modifying the negotiation protocol, we demonstrated that it is possible to produce concessions through use of morality-based utility functions.

5 Related Works

Cooperative negotiation and bargaining problems have been widely studied in the field of multi-agent systems [14, 17, 26, 28]. In the existing studies, if agents cannot reach an agreement, they are forced to concede to each other. Even if agents are assumed to be utility maximizers, the given negotiation protocol forces them to concede and thus they must reduce their own utility [16].

In particular, in the Automated Negotiating Agent Competition (ANAC), participants mainly design concession strategies to find a good agreement [1, 13]. They are assumed to concede under the given negotiation protocol.

Among these current works, few have pursued a morality-based utility function. On the other hand, in utility theory and microeconomics, several works have explored morality-based utility functions, such as the GP model [8] and the DS model [6]. The GP model has many extensions, including a more general model [21] and one changing the order of preferences at different time points [9, 10]. These works can be applied to automated negotiation, but developing such applications will require more time.

6 Conclusion

In this paper, we proposed incorporating a morality-based utility in an automated negotiation model among agents by extending the classic von Neumann and Morgenstern utility models. Then, we analyzed the effect of the morality-based utility model in a concrete example of resource allocation negotiations and demonstrated its characteristics.

Resource allocation negotiation is one of the most discussed topics in automated negotiations. However, in general, an extrinsic negotiation protocol forces agents to concede. It has been rare to find exhaustive discussion on the reasons why agents should have to concede, even if they are utility maximizers. On the other hand, from the results of psychological experiments, we found that there exist mental costs, such as shame, that affect a human's decision making.

In this paper, we analyzed two types of utility functions: a selfish utility function and a morality-based utility function. By applying them to a well-known resource-allocation negotiation problem, we demonstrated the possibility that the morality-based utility function can help agents reach agreement more effectively.

Based on our findings, there are many possible future works. These include extending the two-stage negotiation to an n-stage negotiation and conducting sensitivity analysis on how morality could affect the efficiency of negotiation.

References

1. T. Baarslag, K. Fujita, E. Gerding, K. Hindriks, T. Ito, N.R. Jennings, C. Jonker, S. Kraus, R. Lin, V. Robu, C. Williams, The first international automated negotiating agents competition. Artif. Intel. J. (AIJ) (2012, to appear)
2. J. Steven, Fair division—from cake-cutting to dispute resolution, in *Brams and Alan D* (Cambridge University Press, Taylor, 1996)
3. C.F. Camerer, *Behavioral Game Theory: Experiments in Strategic Interaction. The Roundtable Series in Behavioral Economics* (Princeton University Press, 2003)
4. V. Conitzer, W. Sinnott-Armstrong, J. Schaich Borg, Y. Deng, M. Kramer, Moral decision making frameworks for artificial intelligence, in *Proceedings of Thirty-First AAAI Conference on Artificial Intelligence* (2017)
5. J. Dana, D.M. Cain, R.M. Dawes, What you don't know won't hurt me: costly (but quiet) exit in dictator games. Org. Behav. Human Dec. Process. **100**(2), 193–201 (2006)
6. D. Dillenberger, P. Sadowski, Ashamed to be selfish. Theoret. Econ. **7**(1) (2012)
7. S. Shaheen Fatima, M. Wooldridge, N.R. Jennings, Multi-issue negotiation with deadlines. J. Artif. Intell. Res. (JAIR) **27**, 381–417 (2006)
8. F. Gul, Wolfgang Pesendorfer, Temptation and self-control. Econometrica **69**(6), 1403–1435 (2001)

9. F. Gul, Wolfgang Pesendorfer, Self-control and the theory of consumption. Econometrica **72**(1), 119–158 (2004)
10. F. Gul, W. Pesendorfer, Self control, revealed preferences and consumption choice. Rev. Econ. Dyn. **7**(2), 243–264 (2004)
11. J. Hao, H. Fung Leung, An efficient negotiation protocol to achieve socially optimal allocation, in *PRIMA 2012: Principles and Practice of Multi-Agent Systems*, vol. 7455, Lecture Notes in Computer Science, ed. by Iyad Rahwan, Wayne Wobcke, Sandip Sen, Toshiharu Sugawara (Springer, Berlin, Heidelberg, 2012), pp. 46–60
12. T. Ito, H. Hattori, M. Klein, Multi-issue negotiation protocol for agents: exploring nonlinear utility spaces, in *Proceedings of 20th International Joint Conference on Artificial Intelligence (IJCAI-2007)* (2007), pp. 1347–1352
13. T. Ito, M. Zhang, V. Robu, S. Fatima, T. Matsuo, *Advances in Agent-Based Complex Automated Negotiations* (Springer, 2009)
14. N.R. Jennings, P. Faratin, A.R. Lomuscio, S. Parsons, M. Wooldridge, Carles Sierra, Automated negotiation: prospects, methods, and challenges. Group Decis. Negot. **10**, 199–215 (2001)
15. L.R. Keeney, H. Raiffa, Decisions with multiple objectives, in *Preferences and Value Trade-Offs* (Cambridge University Press, 1993)
16. M. Klein, P. Faratin, H. Sayama, Y. Bar-Yam, Negotiating complex contracts. Group Decis. Negot. **12**(2), 58–73 (2003)
17. S. Kraus, *Strategic Negotiation in Multiagent Environments* (MIT Press, 2001)
18. B.L. Lipman, W. Pesendorfer, Temptation (2011, Working paper)
19. John Nash, The bargaining problem. Econometrica **18**(2), 155–162 (1950)
20. J. Von Neumann, O. Morgenstern, *Theory of Games and Economic Behavior* (Princeton University Press, 1944)
21. J. Noor, N. Takeoka, Uphill self-control. Theoret. Econ. **5**(2), 127–158 (2010)
22. V. Robu, D.J.A. Somefun, J.A. La Poutré, Modeling complex multi-issue negotiations using utility graphs, in*AAMAS '05: Proceedings of the Fourth International Joint Conference on Autonomous Agents and Multiagent Systems* (ACM, New York, NY, USA, 2005), pp. 280–287
23. S.J. Rosenschein, G. Zlotkin, *Rules of Encounter* (MIT Press, 1994)
24. F. Rossi, Moral preferences, in *10th Workshop on Advances in Preference Handling (MPREF)* (2016)
25. S. Saha, S. Sen, An efficient protocol for negotiation over multiple indivisible resources, in *Proceedings of the 20th International Joint Conference on Artifical Intelligence, IJCAI '07* (Morgan Kaufmann Publishers Inc., San Francisco, CA, USA, 2007), pp. 1494–1499
26. T. Sandholm, V.R. Lesser, Issues in automated negotiation and electronic commerce: extending the contract net framework, in *ICMAS1995* (AAAI, 1995), pp. 12–14
27. R.H. Strotz, Myopia and inconsistency in dynamic utility maximiation. Rev. Econ. Stud. **23**, 165–180 (1955)
28. K.P. Sycara, Resolving goal conflicts via negotiation, in *Proceedings of Fifth National Conference on Artificial Intelligence* (1988), pp. 245–250
29. H. Tamura, Y. Nakamura, Decompositions of multiattribute utility functions based on convex dependence. Oper. Res. **31**(3) (1983)

Deniz: A Robust Bidding Strategy for Negotiation Support Systems

Catholijn M. Jonker and Reyhan Aydoğan

Abstract This paper presents the Deniz agent that has been specifically designed to support human negotiators in their bidding. The design of Deniz is done with the criteria of robustness and the availability of small data, due to a small number of negotiation rounds in mind. Deniz's bidding strategy is based on an existing optimal concession strategy that concedes in relation to the expected duration of the negotiation. This accounts for the small data and small number of rounds. Deniz deploys an adaptive behavior-based mechanism to make it robust against exploitation. We tested Deniz against typical bidding strategies and against human negotiators. Our evaluation shows that Deniz is robust against exploitation and gains statistically significant higher utilities than human test subjects, even though it is not designed specifically to get the highest utility against humans.

1 Introduction

Negotiation is part of our daily lives, informally at home, or formally in matters of business. We negotiate to reach a consensus if we have a conflict of interests [7, 12, 15, 20, 21]. While some people are very good at negotiation, others have difficulty in reaching optimal outcomes and mostly end up with suboptimal outcomes [12, 21]. Improving on negotiation outcomes can be done by training the human before they enter the negotiation, or by supporting them during the negotiation. Artificial Intelligence applications have been developed for both purposes. For example, Conflict

C. M. Jonker (✉) · R. Aydoğan
Interactive Intelligence Group Delft University of Technology, Delft, The Netherlands
e-mail: C.M.Jonker@tudelft.nl; c.m.jonker@liacs.leidenuniv.nl

R. Aydoğan
e-mail: reyhan.aydogan@ozyegin.edu.tr

C. M. Jonker
Leiden Institute for Advanced Computer Science, Leiden University, Leiden, The Netherlands

R. Aydoğan
Department of Computer Science, Özyeğin University, Istanbul, Turkey

T. Ito et al. (eds.), *Advances in Automated Negotiations*, Studies in Computational Intelligence 905, https://doi.org/10.1007/978-981-15-5869-6_3

Resolution Agent (CRA) is a virtual agent used to let humans train their negotiation skills [6, 10, 18], and another example can be found in [5].

Decision support systems are meant to support people during their decision making. For negotiation, the Pocket Negotiator is an example [13] that aims at helping human negotiators improve their negotiation outcomes by providing guidance throughout the negotiation process, and with a specialization in bidding support. Another is the Social agent for Advice Provision (SAP) that interacts with the human and attempts to convince her to choose a certain option [2].

In this paper, we focus on agents for negotiation support during the bidding phase. For these agents, the negotiation strategy plays a key role. The literature on automated negotiations has produced a wealth of strategies that have proved themselves in the Automated Negotiating Agents Competition, for an overview see [1, 8, 9, 14]. Even if when focusing only on bidding with complete bids without exchanging any other information with the opponent, these strategies are not directly transferable to human negotiations. The most obvious difference is that in the automated negotiation competitions the agents work with a deadline of some minutes, which is enough for most agents to exchange thousands of bids, whereas in human negotiations the number of rounds is low, although culture-dependent, somewhere between 3 (USA) and 20 (Northern Africa), if you ask the experts. The consequence is that instead of considering big data, in human negotiations we only have small data available to get some inkling about the preferences and the bidding strategy of the opponent. That means that all agents that were developed with elaborate phases for exploring the reaction of the opponent by making random bids are not very suitable for human-human negotiation. Furthermore, we subscribe to the general aim of creating explainable Artificial Intelligence, so the strategy shouldn't be too complex either. This eliminates some more agents. With respect to outcome optimality, for Deniz we chose not only to optimize the utility of Deniz's side in the negotiation, but to optimize outcome in combination with acceptability for human negotiators. For example, in our experience, on average people don't accept bidding advice from a hardheaded agent, as it is too extreme to their taste. This entails that we need to look for a very good concession strategy. Finally, we have the responsibility of creating agents whose strategy is robust against exploitation attempts by the opponent. This means that a straightforward concession strategy is not applicable, but that it should have some aspect of Tit-for-Tat in it. With these criteria in mind, we developed Deniz,[1] an agent that can be used to support humans in their negotiations.

This paper is structured as follows. Section 2 presents the necessary definitions of bids, utilities, and moves, that we need in our description and definition of Deniz. Section 3 defines Deniz, with an emphasis on the explanation of its strategies for determining its next move and a deepening of its concession strategy. The criteria of robustness against exploitation and its suitability for conceding at the right moments are discussed explicitly. The proof of its explainability is in the reading of this section. Section 4 is devoted to the evaluation of Deniz, in which we focus on general performance when supporting humans, and on the criteria of robustness. The

[1] Deniz is a gender-independent name, i.e., it is used for both females and males, and means "sea."

criterion of explainability to the user is left for future work, as it is too much tied into the explainability of the PN framework. We wrap up with our main findings, and conclusions in Sect. 5.

2 Bids, Utilities, and Moves

This section presents the notation and definitions for bids, utilities, and moves as used in the remainder of this article. Furthermore, we would like to point out that our work is inspired on the work by many in the Automated Negotiating Agents Competition (ANAC), as reported in e.g., [14].

Let N be a set of at least two negotiators. For any negotiator $n \in N$, $o_n \in N$ denotes the opponent of n. If no confusion is possible, we drop the index n for the opponent. In the following, our two main negotiators are Deniz agent D, and opponent O.

Let B denotes the space of all possible bids and $b_n^i \in B$ denotes the bid made by negotiator n in round i. Let $u_n : B \mapsto [0, 1]$ denote the utility function of negotiator n. Then, the utility functions of the negotiators map bids $b \in B$ to the two-dimensional utility space $[0, 1] \times [0, 1]$:

$$\langle u_n(b), u_{o_n}(b) \rangle$$

Note that the utility functions of D and O are denoted by u_D and u_O. It is worth noting that in real negotiations, the opponent profiles are typically not known, but estimated. So, in general, the reader should read u_O to be the estimated utility function. For anaylsis sake, we had, of course, the true utility function available.

A move μ_n^i is a tuple (b_n^{i-1}, b_n^i) of two sequential bids of negotiator $n \in N$, made with respect to negotiation round $i > 1$.[2]

For all $i > 1, n, m \in N$, and moves μ_n^i made by n, we define the following:

move size: $\Delta_m(\mu_n^i) = u_m(b_n^i) - u_m(b_n^{i-1})$, is the size of the move (i.e., difference in utility) according to m. As in bilateral negotiations m can be either n or o_n, we consider both Δ_n and Δ_{o_n} for any move.

silent moves: $\mathrm{silent}(\mu_n^i)$ if $|\Delta_n(\mu_n^i)| = 0 \wedge |\Delta_{o_n}(\mu_n^i)| = 0$, which means that agent n made a silent move.

concession moves: $\mathrm{concess}(\mu_n^i)$ if $\Delta_n(\mu_n^i) < 0 \wedge \Delta_{o_n}(\mu_n^i) \geq 0$, which means that negotiator n made a concession.

selfish moves: $\mathrm{selfish}(\mu_n^i)$ if $\Delta_n(\mu_n^i) > 0 \wedge \Delta_{o_n}(\mu_n^i) \leq 0$ which means that n made a selfish move.

nice moves: $\mathrm{nice}(\mu_n^i)$ if $\Delta_n(\mu_n^i) = 0 \wedge \Delta_{o_n}(\mu_n^i) > 0$ which means that n made a nice move, i.e., better for o_n, and for n the utility is the same.

fortune moves: $\mathrm{fort}(\mu_n^i)$ if $\Delta_n(\mu_n^i) > 0 \wedge \Delta_{o_n}(\mu_n^i) > 0$ which means that n made a fortune move, i.e., a move which is better for both negotiators.

---

[2]Note that in the first round, move is undefined.

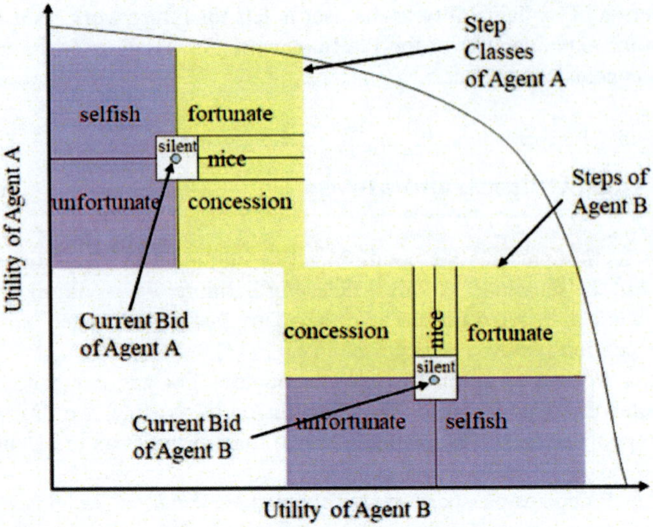

Fig. 1 Moves

unfortunate moves: unfort(μ_n^i) if $\Delta_n(\mu_n^i) \leq 0 \wedge \Delta_{o_n}(\mu_n^i) < 0$, which means that negotiator n made matters worse for both sides.
uncooperative moves: uncoop(μ_n^i) if unfort(μ_n^i) or selfish(μ_n^i).
cooperative moves: coop(μ_n^i) if it is not uncooperative.[3]

These notions are taken from the Dynamics Analysis of Negotiation Strategies (DANS) framework of [11], although the cooperative and uncooperative moves are our additions. Note that in practice it is sometimes useful to take a margin ε around the silent and nice bids as depicted in Fig. 1. The details of defining that precisely are left to the reader.

3 Deniz Strategies

The bidding strategy of the Deniz agent is based on the Greedy Concession Algorithm (GCA) as presented in [3], but with a twist. The twist is that a Tit-for-Tat flavor is added as GCA is to easily exploitable by opponents that recognize that GCA is pure concession strategy. Deniz's negotiation strategy includes a bidding and an acceptance strategy, in which Deniz reasons about the end of the negotiation according to two concepts, i.e., estimated number of negotiation rounds, and negotiation deadline.

The estimated number of negotiation rounds is culture-dependent variable which predicts how many rounds negotiations in a culture typically last. For example, in the

[3]Note that silent moves are thus considered here to be cooperative moves.

USA the number is 3, whereas in Northern Africa the number is somewhere around 20, these numbers we got from talking to negotiation experts. The point for Deniz is not how to define that number, but that it indicates to Deniz when the opponent might lose patience and might end the negotiation without an agreement.

Some negotiations have a deadline; for example some types of auctions, or in negotiations about perishable goods, or in negotiations about transportation, where the departure time of the ship that you would like to give your cargo to is a natural deadline. For Deniz this is important, as it should concede before that deadline (in as far as it is willing to concede). Overall Deniz uses a three-phased negotiation strategy:

Initial phase Deniz initially offers a bid with maximal utility for itself.

Mid phase During most of the negotiation (until last phase), Deniz behaves as explained in Sect. 3.1.

Last phase The last phase starts when there are no rounds left (current round = estimated number of rounds plus 1) or we reached the deadline. Deniz will only make silent bids, and if takes too long to its taste, then it ends the negotiation without an agreement. Too long to its taste is determined by either the deadline, or with a probability that increases with the rounds.

Deniz's acceptance strategy is a thresholded ACNext [4], which means that Deniz will accept an offer if the utility of that offer is greater than or equal to the threshold and the utility of its next bid, if it would make a next bid.

3.1 Making a Next Move

If Deniz decides to make a countermove, then which countermove is decided according to the algorithm defined in this section. In the following, we use M to denote the set of move types {concess, silent, fort, unfort, nice, selfish}, and $(\sigma_i)_{i \in \mathbb{N}}$, with $\sigma_i \in M$ to refer to a sequence of opponent move types in round $i > 1$, e.g., $(\sigma_{i-2}, \sigma_{i-1}, \sigma_i)$. Note that later moves are to the right of earlier moves. Let O^r denote the move type sequence of the last r moves by the Opponent. Then, the next move of Deniz is determined by:

$$next\_move = \begin{cases} same & \text{if } O^3 = (\text{uncoop, uncoop, selfish}) \\ \text{concess} & \text{if } O^3 = (\text{coop, uncoop, selfish}) \\ \text{silent} & \text{if } O^2 = (\text{coop, selfish}) \\ concede\_or\_project & \text{if } O^1 = (\text{unfort}) \\ \text{concess} & \text{otherwise} \end{cases}$$

Note that Deniz is robust against exploitation in human negotiations, as the opponent needs to cooperate every now and then for Deniz to concede. Concessions are done according to the algorithm in Sect. 3.2. The *concede_or_project* procedure is presented in Algorithm 2. To project the opponent's bid to the Pareto Frontier Deniz

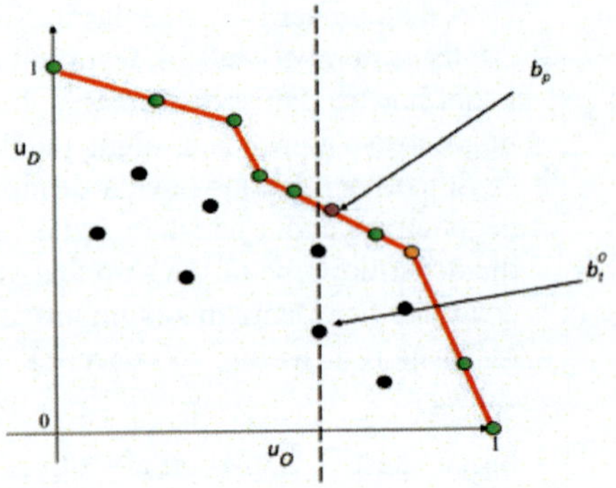

Fig. 2 Projection of the opponent bid to the Pareto frontier

searches for the bid b on the Pareto Frontier, see also Fig. 2, such that $u_{o_n}(b)$ is as close as possible as to $u_{o_n}(b_t^{o_n})$. Here, $b_t^{o_n}$ refers to the last bid made by the opponent, i.e., the bid made in round $t = i - 1$ if Deniz started the negotiation, and $t = i$ if the opponent started the negotiation.[4]

Algorithm 2: Concede or Project to Pareto Procedure

Data: s_c is the current negotiation state, including b_O, which is the opponent's last offer
Result: b is the bid to be offered by our agent

1 $b_c \leftarrow Deniz\_GCA(s_c)$;
2 $b_p \leftarrow ProjectPareto(b_O)$;
3 $b = \text{argmax}\{u_n(b) \mid b \in \{b_c, b_p\}\}$;

3.2 *Concession Strategy*

Deniz first applies GCA, a bidding strategy that determines the optimal concession for a given time t in two steps [3]. Firstly, $f^t(t)$ returns the optimal target conceding utility for that time t. Secondly, it finds a bid with the target utility on the Estimated Pareto Frontier. This gives one bid that is equal to the previous bid or a concession. Deniz improves on this approach by considering this optimal bid as well as nearby bids that may better fit the opponent's preferences. Deniz's next bid determination procedure is defined in Algorithm 3. Note that by explicitly responding to the opponent's last bid, Deniz is a behavior-based variant of GCA. The algorithm uses the following notation:

$b, b' \in B$ are variables over the bid space B, $d \in B$ denotes Deniz's previous bid, and $o \in B$ denotes the last bid by opponent O.

[4] If negotiator n started the negotiation, then in every round i, n is the first to bid, and o_n is the last. So, when in round i, and referring to the last bid made the opponent, it can well be that that bid was made during the previous round.

$s, s_c \in S$: The set of negotiation states S, with variables s, and s_c, where s_c denotes the current negotiation state. A negotiation state refers to the bid history (of bids made so far, annotated by who made which bid), the deadline, current time, and the discount factor (if a discount factor is used).
f^t: $S \mapsto [0, 1]$: GCA's function to determine the target utility, on the basis of a negotiation state, see [3].
E: the Estimated Pareto Front, i.e., the Pareto Front corresponding to u_D and u_O. As u_O is typically estimated, also E is an estimation.
h: $BxB \mapsto \mathbb{N}$: the hamming distance between bid b and b', defined as the number of bid issues that have different values in b and b'.
$\mathcal{R}$: $\mathcal{P}(B) \mapsto B$ is a function that randomly picks an element from a set, here applied to B.

Algorithm 3: Deniz GCA Concession Algorithm

Data: s_c is the current negotiation state
Result: b_c is the bid to be offered by our agent

```
/* Determine the target utility for the optimal bid */
tu ← f^t(s_c);
/* Determine B_tu as the bids on the Estimated Pareto Frontier
that are closest to the optimal target utility for Deniz.
Note that in discrete domains, more than one bid in E might
exist that satisfies the constraint */
B_tu ← {b ∈ E | argmin_b |u_D(b) − tu|};
/* tu' is the adapted target utility */
tu' ← u_D(b), where b ∈ B_tu;
/* Create the set C of all possible concessions between the
target utility and Deniz's previous bid d. As a concession
the bids should have an opponent utility that is higher or
equal than that of Deniz's previous bid */
C ← {b ∈ B | tu' ≤ u_D(b) < u_D(d) ∧ u_O(d) ≤ u_O(b)};
/* If C is empty, then we fall back to bids that have Deniz's
previous bid utility. This set is non-empty, as it always
contains Deniz's previous bid d */
if C = ∅ then
  | C ← {b ∈ B | u_D(b) = u_D(d)};
end
/* Return a randomly chosen bid from the bids in C that have
the smallest Hamming distance to the opponent's last bid */
b_c ← R({b ∈ C | argmin_b h(b, o)});
```

4 Evaluation

We evaluated Deniz by letting it play against itself, a hardheaded opponent, an explorative agent, and against agents playing some concession strategy. Furthermore, we experimented with human participants that we asked to play against Deniz and then asked about their experience. In all cases, we used the Job domain, with default settings for the preferences of employer and employee. The number of expected rounds was set to 10 in all experiments. This number is used by Deniz in its bidding strategy (to know when to concede) and in its acceptance strategy (to know when to accept an acceptable bid). Also, we gave Deniz a internal (private) deadline of 15 min, after which Deniz will end the negotiation without an agreement.

4.1 Job Domain

The job domain is that of an prospective young ICT professionals negotiating with his/her prospective employer about the job conditions.

The profiles for this domain were obtained in 2010 by interviewing young ICT professionals and HR officers from ICT companies [19]. This resulted in the issue descriptions provided in Table 1, in which fte stands for full time equivalent:

Furthermore, we asked all participants about their preferences, and on the basis of that we set up a typical profile for the employees and one for the employer side, and created utility functions for them. For this paper, the exact procedure is not important, nor is the representativeness of the profiles. Note that the domain size is small. To be precise that is 540, if the salary options are done in steps of 500 Euro.

We also compared the outcomes of the negotiations to the Nash Product, which is the maximum of the product of the utilities over all possible bids. In general, for a set of utility functions U, and a bid space B the Nash Product $\eta(U, B)$ is defined as:

$$\eta(U, B) = \max\{\prod_{u \in U} u(b) | b \in B\}$$

Table 1 Issues with value range of the job domain

Issue	Value range	Employee profile	Employer profile
Salary	[2000–4000 Euro]	Higher is better	Lower is better
Fte	0.6, 0.8, 1.0	$1.0 > 0.8\ 0.6$	$1.0 > 0.8\ 0.6$
Work from home	0, 1, 2 days	$2 > 1 > 0$	$0 > 1 > 0$
Lease car	Yes no	Yes > No	Yes < No
Permanent contract	Yes no	Yes > No	Yes < No
Career opportunities	Low, medium, high	High > medium > low	High < medium < low

Table 2 The second column shows the agreement reached when both employer and employee were played by Deniz. The third column shows the bid corresponding to the nash product

Issue	Deniz-Deniz	Nash product bid
Salary	3000	4000
Fte	1.0	1.0
Work from home	0	0
Lease car	No	No
Permanent contract	Yes	Yes
Career development opportunities	Low	Low
(*Employer utility, employee utility*)	(0.68, 0.58)	(0.58, 0.70)

Similarly, we define $\beta(U, B)$ to a Nash Product bid, i.e., a bid in B such that its utilities form a Nash Product with respect to the set of utilities U. If no confusion is expected as U and B are fixed, we simply write η and β. In the Job domain, there is only one Nash Product bid, which is described in Table 2.

4.2 Deniz Against Other Negotiating Agents

Deniz agent when playing itself gets the result depicted in Fig. 3 and elaborated in Table 2. As experimenter we played the Employee role and for every bid asked a recommendation of Deniz, which we then offered without any changes to the opponent (played by a copy of Deniz). The negotiation lasted 10 rounds.

We also negotiated against Deniz ourselves and to test if Deniz takes advantage of conceding opponents, we first played a negotiator that always concedes. As you can see in Fig. 4, part (a), Deniz also concedes, but follows the Greedy Concession Algorithm, which concedes much slower. To prevent a possible failure of negotiation, we accepted the offer by Deniz in round 10, which was the estimated deadline for the negotiation.

Then, we played a very reasonable opponent that starts its bidding close to the Nash Product and is trying to reach a fair outcome, see part (b) of Fig. 4. In our offers, we stayed close to what we considered fair and waited for Deniz to concede to us. The bids we offered had the following utility pattern with respect to our utility function was: 0.7, 0.72, 0.64, 0.7, 0.58, 0.64, 0.72, 0.58, 0.54, 0.58 and Deniz accepted our last offer (in round 10).

Finally, we played a somewhat explorative strategy, see part (c) of Fig. 4. In that negotiation, we deliberately sometimes offered bids that are below the Pareto Optimal Frontier and also sometimes conceded and then went back up again. As can be seen, Deniz is not confused by this and slowly concedes after its opponent makes

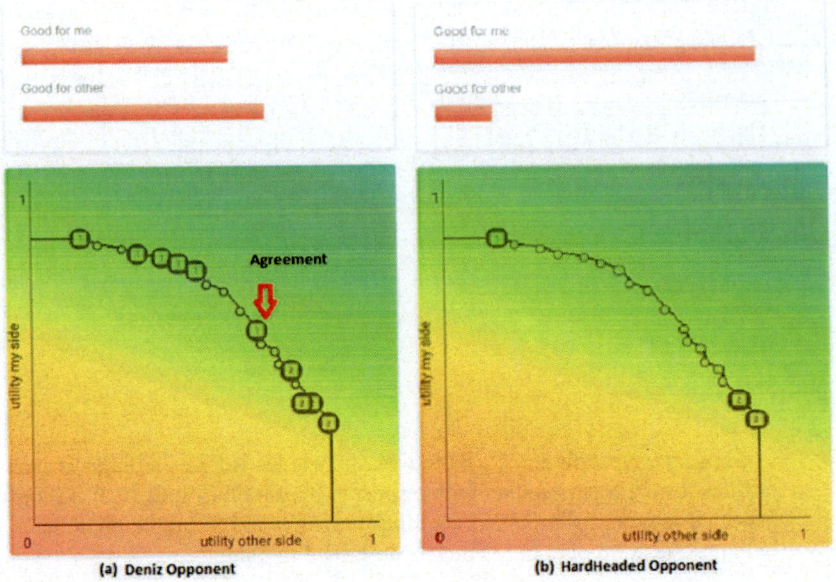

Fig. 3 Negotiation dance of Deniz versus Deniz (on the left) and Deniz versus HardHeaded (on the right) in the jobs domain. Note that no agreement was reached here

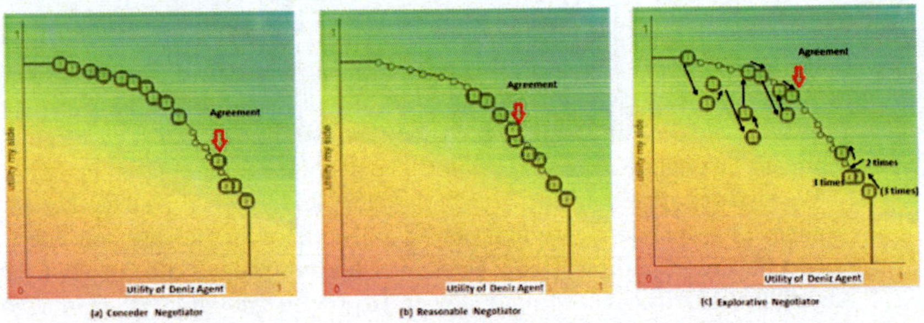

Fig. 4 Negotiation Dance of Deniz against **a** conceder, **b** reasonable, and **c** explorative human negotiator in the jobs domain, The agreement configurations and utilities are presented in Table 3

a concession. Furthermore, it also shows that Deniz agent will accept a reasonable offer when the estimated deadline is reached.

Note that Deniz can only make bids on the Pareto Optimal frontier if the estimation of the utility function of its opponent is correct. In these experiments, Deniz (and both parties) always obtained the correct utility functions from the authors. Note that the depicted Pareto Frontier is one that is computed by PN.

Table 3 Agreements of the negotiations of Deniz against different opponents

Issue	Conceder	Reasonable	Explorative
Salary	2000	3000	4000
Fte	1.0	1.0	1.0
Work from home	0	0	0
Lease car	No	No	No
Permanent contract	Yes	Yes	Yes
Career development opportunities	Low	Low	Medium
(u_D, u_O)	(0.78, 0.46)	(0.68, 0,58)	(0.53, 0.72)

4.3 Deniz Negotiating Against Humans

In the previous section, we as authors of Deniz tried to exploit Deniz strategy and found it to be robust against our efforts. We could get some results (0.7 utility) if we did our best, and we had the help of PN overview of the bid space, including the Pareto Optimal Frontier, see Fig. 5. In this section, we present and discuss the results of an experiment in which humans that were unfamiliar with the Deniz agent negotiate against Deniz. The humans were asked to play against Deniz in the PN framework, which makes it possible to toggle between a supported and an unsupported version. The experiment was approved by TU Delft's Human Research Ethics committee, the data of participants that gave informed consent, is stored according to the Data Management Plan approved by TU Delft's Data Steward.

In PN bidding support interface, the middle section allows users to put in their own choices per issue. The red bars in the right part indicate for a selected point i the bid space, how good that is from the perspectives of the user and the opponent. Note that this is done using the estimated opponent profile that can be constructed in another part of PN. The graph on the right shows an overview of the bid space with the Estimated Pareto Optimal Frontier, and all the bids made so far. Note that the user can click to select points on the Pareto Optimal Frontier. If done so, the selected bid's content is copied to the middle section of interface. Finally, note that at the bottom, the user can ask PN for a bid suggestion, and can also accept a bid from the opponent or walk away without an agreement by ending the negotiation.

In the unsupported version of PN, the users only have the middle section available, where they can enter their offers. The red bars, the graph, and the button to ask for a suggestion are not available.

For our experiment, we gathered 78 participants from three classes of students. The first group consisted of Computer Science students and the second group of Industrial Engineering students. These groups of students studied at Özyeğin University (Turkey). The third group consists of business students at Erasmus University (the Netherlands). We did not inform the participants that Deniz agent was also their supporting agent in the condition that they received support from PN.

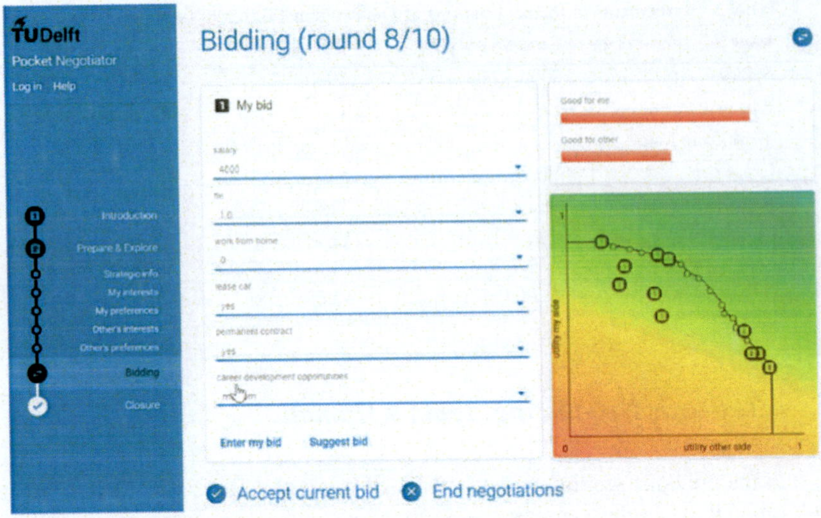

Fig. 5 Pocket negotiator bidding interface

Table 4 Deniz against human negotiators: statistical analysis

	No support		Support		All	
	Human	Deniz	Human	Deniz	Human	Deniz
Average utility	0.50	0.69	0.53	0.68	0.50	0.66
Standard deviation	0.09	0.10	0.09	0.09	0.13	0.15
t value	9.50		7.71		12.16	
p value	<0.00001		<0.00001		<0.00001	

The average utility received by the participants when negotiating against Deniz without support from PN was 0.50, whereas with the help from PN they received an average utility of 0.53. In the support condition, they had the option of asking for a recommendation on what to bid. The agent that would give them that advice was also Deniz. However, on average the participants only asked for a recommendation 2.3 times, and 29 participants never asked for a recommendation at all. Deniz agent on the other hand, in the no-support condition had an average of 0.69 utility, and in the support condition on average 0.68 utility (Std = 0.09). In all conditions, and also when considering all negotiation sessions together, Deniz had a statistically significant higher average utility than the human participants, see Table 4.

We asked the participant how they felt about Deniz as an opponent through the following statements using a 5-point Likert scale:

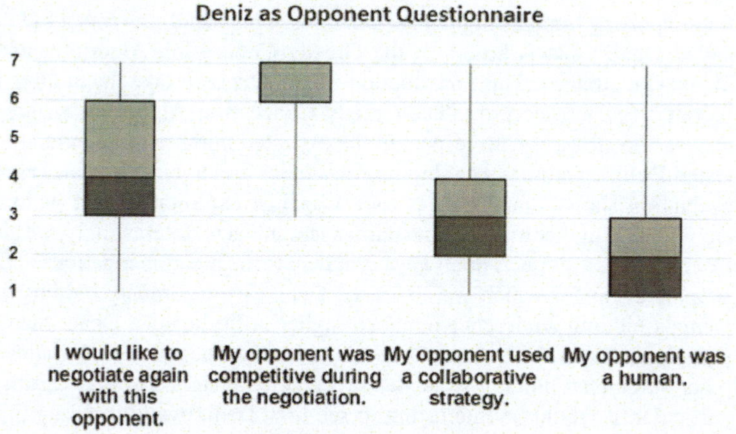

Fig. 6 Participants thoughts about Deniz agent as an opponent

1. I would like to negotiate again with this opponent sometime in the future.
2. My opponent was competitive during the negotiation.
3. My opponent used a collaborative strategy.
4. My opponent was a human.

The results are presented in Fig. 6. The participants clearly found Deniz to be a competitive agent that did not seem human-like to them. The responses to the statement "I would like to negotiate again with this opponent" gives us more information on the differences between the groups, which was statistically significant $p = 0.017$; $F\text{ratio} = 4.4$ and $\mathrm{d}F = 2$. The averages per group are as follows Group 1: 4.9; Group 2: 4.9, and Group 3: 3.4. Given that the Deniz agent strategy is a bit like Tit-for-Tat, meaning that non-conceding behavior would result in no concessions by Deniz, and the fact that Group 3 participant are trained in not giving in too easily, they would find Deniz more tough to play against than the other groups and thus they would be less eager to play Deniz again. This can be seen from the analysis of the types of moves made by the different groups, see Table ?. In particular, Table ? shows that in Group 3 the participants (P) make more silent moves than the participants from the other groups. This causes the opponent (Deniz agent, marked by D) to also make more silent moves.

5 Conclusion

The literature on automated negotiation has produced a wealth of automated bidding strategies that have proved themselves in many negotiation domains and against many automated negotiating agents. However, these agents have not been designed for supporting humans in their negotiations. This paper presents a bidding strategy, called Deniz, that is based on the Greedy Concession Algorithm and deploys a Tit-for-Tat strategy. This construction allows Deniz to take the bidding behavior of its opponent into account. The Greedy Concession Algorithm ensure an optimal concession strategy with respect to the expected number of negotiation rounds. We tested Deniz against typical bidding strategies and against human negotiators. Our evaluation shows that Deniz is robust against exploitation and gains statistically significantly higher utilities than human test subjects. Interestingly, we also observed that human test subjects don't always adhere to the recommendations of their bidding support agent for which we also used Deniz. By rigorously bidding according to Deniz's bidding advice, we obtained higher utility against Deniz than the average participant. Deniz agent should be tested in other negotiation domains and against other automated bidding strategies specifically designed to play against human. In particular, it would be interesting to see how Deniz would hold up in comparison with the agents presented in [16, 17]. However, the agents do not run in the same platform, so this requires re-implementation of these agents and was therefore left for future work. Furthermore, it would be important to develop a host of bidding support strategies that match with different human negotiation styles.

Further directions of research are to combine our bidding support strategies with other work, such as that of Gratch and co-authors, on training people to become better negotiators.

6 Preference Profiles

See Tables 5 and 6.

Table 5 Importance of each issue for employer and employee

Weights of issues	Employer's preferences	Employee preferences
Salary	0.2	0.24
Fte	0.165	0.32
Work from home	0.245	0.18
Lease car	0.08	0.06
Permanent contract	0.21	0.16
Career development opportunity	0.10	0.04

Table 6 Evaluation values of employer and employee

Issue		Employer's preferences	Employee preferences
Salary	Value	Evaluation value	Evaluation value
	2000	1.0	0.0
	2500	0.75	0.25
	3000	0.5	0.50
	3500	0.25	0.75
	4000	0.0	1.0
Fte	Value	Evaluation value	Evaluation value
	0.6	0.36	0.25
	0.8	0.5	0.50
	1.0	0.64	0.75
Work from home	Value	Evaluation value	Evaluation value
	0 day	1.0	0.33
	1 day	0.5	0.50
	2 days	0.0	0.67
Lease car	Value	Evaluation value	Evaluation value
	Yes	0.0	1.0
	No	1.0	0.0
Permanent contract	Value	Evaluation value	Evaluation value
	Yes	0.24	1.0
	No	0.76	0.0
Career opportunity	Value	Evaluation value	Evaluation value
	Low	1.0	0.0
	Medium	0.5	0.5
	High	0.0	1.0

Acknowledgements We thank all participants that over the years helped us in our experiments. We also give specific thanks to scientific programmer Wouter Pasman, technician Bart Vastenhouw, Dimitrios Teskouras, professor of the course at Erasmus University that hosted part of our experiments, and our collaborator in many negotiation research projects Tim Baarslag.

References

1. R. Aydoğan, T. Baarslag, C.M. Jonker, K. Fujita, T. Ito, R. Hadfi, K. Hayakawa, A baseline for non-linear bilateral negotiations: the full results of the agents competing in anac 2014 (2016)
2. A. Azaria, Z. Rabinovich, S. Kraus, C.V. Goldman, Y. Gal, Strategic advice provision in repeated human-agent interactions. Inst. Adv. Comput. Stud. Univ. Maryl. **1500**(20742) (2012)
3. T. Baarslag, E.H. Gerding, R. Aydoğan, M.C. Schraefel, Optimal negotiation decision functions in time-sensitive domains, in *2015 IEEE/WIC/ACM International Conference on Web Intelligence and Intelligent Agent Technology (WI-IAT)*, vol. 2 (2015), pp. 190–197

4. T. Baarslag, K.V. Hindriks, C.M. Jonker, Effective acceptance conditions in real-time automated negotiation. Dec. Support Syst. **60**, 68–77 (2014). Apr
5. J. Broekens, M. Harbers, W.-P. Brinkman, C.M. Jonker, K. Van den Bosch, J.-J. Meyer, Virtual reality negotiation training increases negotiation knowledge and skill, in *International Conference on Intelligent Virtual Agents* (Springer, 2012), pp. 218–230
6. D. DeVault, J. Mell, J. Gratch, Toward natural turn-taking in a virtual human negotiation agent, in *2015 AAAI Spring Symposium Series* (2015)
7. R. Fisher, W.L. Ury, B. Patton, Negotiating agreement without giving in. Penguin, Getting to yes (2011)
8. K. Fujita, R. Aydoğan, T. Baarslag, K. Hindriks, T. Ito, C. Jonker, The sixth automated negotiating agents competition (anac 2015), in *Modern Approaches to Agent-based Complex Automated Negotiation* (Springer, 2017), pp. 139–151
9. K. Fujita, R. Aydoğan, T. Baarslag, T. Ito, C. Jonker, The fifth automated negotiating agents competition (anac 2014), in *Recent Advances in Agent-based Complex Automated Negotiation* (Springer, 2016), pp. 211–224
10. J. Gratch, D. DeVault, G. Lucas, The benefits of virtual humans for teaching negotiation, in *Proceedings of the 16th International Conference on Intelligent Virtual Agents (IVA), September 2016* (Springer, Los Angeles, CA, 2016)
11. K. Hindriks, C.M. Jonker, D. Tykhonov, Let's dans! an analytic framework of negotiation dynamics and strategies. Web Intel. Agent Syst. Int. J. **9**(4), 319–335 (2011)
12. J. Richardson Howard Raiffa, D. Metcalfe, *Negotiation Analysis: The Science and Art of Collaborative Decision Making* (2002)
13. C.M. Jonker, R. Aydoğan, T. Baarslag, J. Broekens, C.A. Detweiler, K.V. Hindriks, A. Huldtgren, W. Pasman, An introduction to the pocket negotiator: a general purpose negotiation support system, in *Multi-Agent Systems and Agreement Technologies* (Springer, 2016), pp. 13–27
14. C.M. Jonker, R. Aydoğan, T. Baarslag, K. Fujita, T. Ito, K.V. Hindriks, Automated negotiating agents competition (ANAC). AAAI, 5070–5072 (2017
15. R.J. Lewicki, D.M. Saunders, B. Barry, John W. Minton, *Essentials of Negotiation* (McGraw-Hill, Boston, MA, 2003)
16. R. Lin, S. Kraus, Can automated agents proficiently negotiate with humans? Commun. ACM **53**(1), 78–88 (2010)
17. R. Lin, Y. Oshrat, S. Kraus, Automated agents that proficiently negotiate with people: can we keep people out of the evaluation loop, in *New Trends in Agent-Based Complex Automated Negotiations* (Springer, 2012), pp. 57–80
18. J. Mell, J. Gratch, T. Baarslag, R. Aydoğan, C.M. Jonker, Results of the first annual human-agent league of the automated negotiating agents competition, in *Proceedings of the 18th International Conference on Intelligent Virtual Agents* (ACM, 2018), pp. 23–28
19. A. Pommeranz, W. Visser, J. Broekens, P. Wiggers, K. Hindriks, C.M. Jonker, Duo meta-model for knowledge elicitation and bidding support in nss, in *Proceedings of 11th Group Decision and Negotiation Conference* (2010), pp. 120–123
20. Howard Raiffa, *The Art and Science of Negotiation: How to Resolve Conflicts and Get the Best Out of Bargaining* (Harvard University Press, Cambridge, MA, 1982)
21. L. Thompson, *The Mind and Heart of the Negotiator*, 3rd edn. (Prentice Hall Press, Upper Saddle River, NJ, USA, 2000)

Collaborative Privacy Management with Auctioning Mechanisms

Onuralp Ulusoy and Pınar Yolum

Abstract Online social networks enable users to share content with other users. Many times, a shared content, such as a group picture, may reveal private information about the uploader as well as others who are associated with the content. Ideally, protection of privacy in such cases would need to consider the privacy concerns of all relevant individuals. However, these concerns might conflict and satisfying one user's privacy needs could cause a privacy violation for others. This calls for computational mechanisms that can decide on the privacy policies of the content collaboratively. Accordingly, we propose an agent-based collaborative privacy management model for online social networks (OSNs). Agents represent OSN users and manage their privacy requirements on their behalf. We extend Clarke-Tax mechanism for auctioning to achieve fair handling of privacy settings and to tax the agents whose privacy settings are chosen. We evaluate our approach over multi-agent simulations and show that it produces privacy policies efficiently and more accurately than existing approaches.

1 Introduction

Privacy is one of the most important notions that should be taken into consideration while using online social networks (OSNs). Even a seemingly unimportant content sharing might have drastic effects on the life of the publisher, or in some cases, other people that are related to the content. With the tremendous worldwide usage of the OSNs, these violations might occur frequently. Hence, privacy resolution mechanisms are essential, especially for the published contents that concern multiple

O. Ulusoy (✉) · P. Yolum
Department of Information and Computing Sciences, Utrecht University, Utrecht, The Netherlands
e-mail: o.ulusoy@uu.nl

P. Yolum
e-mail: p.yolum@uu.nl

T. Ito et al. (eds.), *Advances in Automated Negotiations*, Studies in Computational Intelligence 905, https://doi.org/10.1007/978-981-15-5869-6_4

people. Collaborative privacy management regulate such settings by enabling all the affected users to have their say in publishing or not publishing a content.

Reaching a consensus in collaborative privacy management is usually not an easy task, since people's privacy requirements can easily be in conflict. In real-life cases, the resolution requires much time and effort through mechanisms like negotiation, argumentation, and so on. However, current most widely used OSNs only enable their users to tweak with their own privacy settings while publishing contents. Since they can't provide collaborative solutions, people tend to resolve conflicts via different media [6] or in most cases, ignore others' privacy requirements and cause voluntary or involuntary privacy violations. In order to be able to handle them online, decision mechanisms should be in place. Recently, multi-agent agreement techniques have been used to address collaborative privacy management. Kekulluoglu et al. [4] and Such and Rovatsos [11] propose negotiation-based approaches that enable users to reach a consensus on how to share a content. Kokciyan et al. [5] use argumentation to enable one user to persuade the other into sharing with her own privacy constraints. These approaches have been successful but require heavy computations; that is, they can only be used when the entities can reason on its privacy policies and communicate with others intensively.

On a different line, Squicciarini et al. [9] propose a model where users enter auctions for deciding on a policy that requires collaborative management over a content. Each user creates bids based on how much she wants to see a content public or private. In that approach, users collect currencies by publishing content and tagging people that are related to content. These currencies are used in an auction, where users spend their currencies to convince other users to accept their policy, based on Clarke-Tax mechanism [1, 2]. However, with its current state, the system is open to abuse by the users, such that a single user's privacy can be ignored repeatedly when all others collaborate strongly. Further, since the bids are expected to be generated by users individually for each post, it is difficult to apply them in real-life online social networks.

Ensuring collaborative privacy management in a real-life scale OSN requires the system to scale to the tremendous amount of content being shared. To enable this, first, the operations expected from users should be handled automatically so that users do not need to think through the operations for each content. Second, the proposed mechanism should be easy to compute, because it will be repeated for each content separately. Third and most importantly, it should preserve privacy of the users fairly so that no user is left at a disadvantage.

Accordingly, this paper proposes PANO an agent-based collaborative privacy management system that uses ideas from the work proposed by Squicciarini et al. [9]. There are three main contributions of PANO: First, it employs agents for privacy management, where agents act on behalf of users to enforce their privacy constraints, so that heavy user involvement is reduced to minimum. The agents manage their users' privacy constraints and bid on behalf of them. Second, it contains a fair reward mechanism, which is protective against abuses, and at the same time encourages users to share content online. Third, it works with a group-wise currency system in auctions, where the agents cannot use the advantages they gain from the system against individuals. This disables an agent to abuse another agent's privacy. Our experimental evaluation shows that agents can indeed carry out this task successfully and help preserve their users' privacy with high accuracy.

The rest of this paper is organized as follows: Sect. 2.1 gives the necessary background for Clarke-Tax mechanism. Section 2.2 explains our agent-based approach PANO to handle privacy management. Section 3 presents the agent bidding mechanism and the proposed policy, bidding and robustness strategies to construct PANO. Section 4 explains the rules and process of the simulation implemented to evaluate our approach. Section 5 evaluates the success of PANO. Section 6 discusses our work in relation to existing methods in the literature and gives pointers for future work.

2 Auctioning Privacy

As a broad definition, privacy is the concept of individuals deciding on how much about themselves to be shared with the others. In OSNs, these decisions can be represented with privacy policies. Applying privacy policies when the information is solely related to an individual itself is an easy task, when the necessary tools are provided. However, a piece of information, e.g., a photograph content in an OSN, can be related to more than one individual. In such cases, the decisions of the individuals for the extend of how much to share may differ, resulting in conflicts. These conflicts require some resolution mechanism to define a generalized privacy policy with the goal to comply with every individual's privacy requirements. For this conflict resolution mechanism, we propose PANO, which employs Clarke-Tax mechanism with an agent-based approach.

2.1 Background: Clarke-Tax Mechanism

Clarke-Tax mechanism [1] provides an auction mechanism, similar to English auctions where participants bid for different, possible actions in the environment. The action that receives the highest total bids from the participants wins and is executed. Different from an English auction, participants who aid in the winning action to be chosen, i.e., that bid toward it, are taxed according to the value they put on it. This is achieved by subtracting the bid values of every single user from the overall values. If the subtraction of a single user's bid changes the overall decision, it shows that the user's bid on this action had a *decisive* value. Thus, the user is taxed with the difference of the actual action's score and the score of action to be taken if that user was not present in the auction.

In the context of collaborative privacy, Clarke-Tax mechanism is used to decide on how an image is going to be shared. Squicciarini et al. [9] consider three types of sharing actions: *no share*, *limited share*, and *public share*. We follow the same scheme here. When an image is about to be shared, all the relevant participants bid on

Table 1 Four user bids for sharing an image

Users	No share	Limited share	Public share
Alice	3	5	0
Bob	15	2	0
Carol	5	8	5
Dave	2	6	18

Table 2 Clarke-Tax mechanism example—decision and taxes

Values	No	Limited	Public	Taxes
Overall	**25**	21	23	–
Without Alice	22	16	**23**	1
Without Bob	10	19	**23**	13
Without Carol	**20**	13	18	0
Without Dave	**23**	15	5	0

these three possible actions. Table 1 shows an example of biddings of four users for deciding to share or not share a content. Users decide based on their own importance of the three actions. According to Table 1, it can be seen that Bob values the *No Share* action more than the others, while Dave values *Public Share* action the most.

According to the biddings of all users, Clarke-Tax auction mechanism decides on which action to take. Based on the bids from Table 1, no share action receives a total of 25, whereas limited share receives a total of 21, and public share a total of 23 points. Therefore, the no share action is chosen. Table 2 shows the resulting decision and applies the taxes according to the biddings in Table 1. According to the decided action, each user that bid for the decisive action is taxed.

Table 2 shows that Alice and Bob are taxed, since each of these users' absence in the auction causes the decisive action to be changed. When the scores of Alice is subtracted from the overall score, the decision of sharing the content receives the maximum score by 23, while not sharing gets a score of 22. This causes Alice to be taxed with a score of 1. As mentioned above, Bob bid a greater value for not sharing, and its absence from the auction also causes the action to be changed. Since the differences of the actions are much bigger in Bob's case (i.e., 13, obtained from the subtraction of *Public Share* and *No Share* scores), Bob is taxed with a much greater value. It is important to note that although the user is taxed, he gets the action to be decided what it values most, and prevents the content from being shared. The importance of good evaluation and truthfulness for bidding are crucial. For example, if Bob bid for not sharing with a rather big value, even though the decision was not that important to him, it would have paid the bid amount plus a great amount of tax. On the contrary, if he had bid for much less, then his privacy might have been violated. Hence, it is important to be able to create bids that reflect the true evaluations of the users.

2.2 Challenges

Auctioning with Clarke-Tax Mechanism is an efficient way of negotiation, since it has been shown that truthfulness is the best strategy for bidding [9]. The bidder who overvalues a decision to get its way can be taxed with a greater amount, because it changes the group decision by spending system currency way more than the other participants of the auction. This results in the participants bidding with truthful values, while trying to establish its own decision and not get taxed with a greater amount. Even with the notion of truthfulness, applying a pure Clarke-Tax approach still has some limitations that can result in abuse by the bidders or inflation in the currency used. Consider the following examples:

Example 1 Dave wants to share a photograph in which he is with Alice, Bob, and Carol. Alice and Carol do not want the picture to be shared. Dave has previously shared a lot of photographs that are unrelated to Alice, Bob, and Carol and gained substantially more currencies than the others. Since Dave can spend more than the others, he puts a high bid (e.g. >50 for the bids in Table 1), instead of the 18 that was bid for sharing in Table 1, to share the photograph, which bids of Alice, Bob and Carol cannot match. Therefore, the photograph is shared in the social network even though the majority of the people in the photograph did not want it to be shared.

Another drawback of the application of Clarke-Tax mechanism over OSNs is that it requires user involvement for every single auction. This could be necessary for some cases, but it could become a tedious task for greater number of contents. Also, unavailability of some users in auctions who are related to a content that has privacy conflicts could make the method get stuck, or decide on a semi-successful policy. To resolve these issues, an automated auction process where software agents represent the users and act on their behalf can be implemented.

Example 2 Alice, Bob, Carol, and Dave want to decide on sharing or not sharing a photograph they are in, in a social network. Carol shares a lot of photographs daily, has a lot of currency to spend, and does not want to join a Clarke-Tax auction, but still wants the photograph to be shared. Alice also doesn't mind if the photograph is shared or not, and doesn't enter the auction. Bob and Dave have a disagreement, and join into an auction, and Bob wins the auction for not sharing the photograph, by outbidding Dave. The final policy is decided for not sharing the photograph, even though majority of the people in the photograph wanted it to be shared.

3 Agent-Based Bidding

The pure Clark-Tax-based mechanism in Squicciarini et al. [9] requires user involvement for auctions. This could become a tedious work for the OSN users that shares a multitude of contents every day. Thus, we develop an agent-based approach, where each user is represented with an agent that maintains its user's privacy constraints, manages total currencies, and generates bids when necessary. In principle, understanding users' privacy constraints automatically is difficult. It would require the user behavior to be modeled and privacy constraints to be learned over time. There is a good body of literature on learning privacy constraints [8, 10]. Here, we assume that the user's agent is already aware of the constraints, either through learning or through elicitation.

3.1 Privacy Policy

Clarke-Tax auction mechanism depends on the decisions of the participants. These participants have a self-contained evaluation calculation to decide on the importance of different actions, and bids according to the result of the evaluation. If an agent can assess the importance of a content for its user correctly, and bid with neither excessive nor low amounts, it can help to get better decisive actions, while preserving the previously obtained scores of the user within the network.

PANO makes use of policies for the agents to compute the bidding evaluations. Agents have multiple policies that correspond to different actions, and in an auction, they correspond to these policies to place bids accordingly. In PANO, a policy P is a 5-tuple P = *{a, n, p, q, i}*, where a is the agent that the policy belongs to, n is the set of users in the network the policy is applied to, p is the conditions for the content types that the policy will be applied, q is the action of sharing or not sharing the contents when the policy is applied, and i is the importance of the policy, which is a value between 0 and 1. An example policy of Alice wanting to share photographs that contain scenery tag with friends, with 0.6 importance can be represented as P = *{Alice, friends, photograph[scenery], share, 0.6}*.

The success of the mechanism depends on how the final policy out of an auction satisfies the policies of the agents. The resulting policy of an auction should correctly assign the policy-applicable users of the network where there are no conflicts between the auction participants, and try to assign the rest as satisfactory as possible to protect the common good. Equation 1 measures how well the overall result found with PANO satisfies the n agents that enter the auction. Success is defined as the number of the users that the applied policy differing from the agents requirements (UPC: count of the users with unsatisfied policy for user u), divided by the entire set of users that were considered to share the content with (TNU: total count of users in auction participants' network).

$$\text{Success\%} = \left(1 - \frac{\sum_{u=1}^{n} UPC}{TNU}\right) * 100 \tag{1}$$

Example 3 Consider two agents in a network of 200 agents that enter an auction to decide how to share a picture. The first agent wants to share the picture with 140 people, and the second agents wants to share it with all. Assume that as a result, the picture is shared with 160 users that include the 140 users that the first agent preferred. The metric would result in $(1 - (20 + 40)/200) * 100 = 70\%$ success.

With the given policy notation, satisfaction of individual users can also be calculated. Equation 2 measures the user satisfaction after an auction, considering how well the outcome is aligned with the agent's policy and the importance of the policy. That is, while the satisfaction value for a single content can be computed with the Success metric, making use of importance values of policies can give us sensitivity levels (SL) of users for conditions of content types that are also represented in the policies. For example, if a user has importance level of 0.6 for a policy that is related to a condition of a content type, it can be assumed that the sensitivity level of the user for the same condition can also be given as 0.6. For multiple policies of the same user for the same condition, we take the average of importance levels of the related policies for SL value. Using the satisfaction metric for a single content as CS, and the sensitivity level of the content for the user as SL, we define the user satisfaction (US) metric for an agent with the formula below, where i is the content id from the previously policy applied contents.

$$US = \sum_{i=1}^{n}(SL_i * CS_i) / \sum_{i=1}^{n}(SL_i) \tag{2}$$

Example 4 Alice wants to share a photograph tagged with "bar," which also has other co-owners. She has a policy P for contents that are tagged with "bar," and policy P indicates Alice preferring contents tagged with "bar" to be shared with her "friends," with a sensitivity level of 0.7. According to the final policy decided with Clarke-Tax mechanism, the content is shared with 70 of her friends, while her policy was intended to share it with 100 people. The satisfaction of Alice for this content is 0.7. If Alice have shared a content before with a satisfaction of 0.6 where her policy had sensitivity level of 1.0, the combined user satisfaction is calculated with the user satisfaction metric as:

$$US_{Alice} = (0.6 * 1.0 + 0.7 * 0.7)/(1.0 + 0.7) = 0.64.$$

3.2 Preventing Abuses in the Auction Mechanism

The Clarke-Tax auctions are beneficial for decision making for multiple participants with different opinions, as they support truthfulness. However, the economic system and the currency used in the mechanism can allow abuses, as explained in Sect. 2.2. In order to prevent the system from facing malicious behavior by some users, some modifications are need to be made for earning the currency and spending it. The main modifications proposed in this paper to prevent abuses are the group-wise bid scoring, boundaries of the bids for the auctions and the balance between the income and expends.

Group-wise Spending: To prevent abuse of using currencies for irrelevant auctions with different users, earned currencies can only be used in new contents with same co-owners. Recall that there are three types of actions: not sharing with anyone, sharing with a limited number of people, and sharing with public. Limited number of people is decided from the conflicted share decisions of users. For example, when a co-owner of the content decides not to share it with a specific user and another co-owner wants to share the content with it, this user is added to the limited audience list. Since the conflicts, and the audience lists for actions are only related to the policies of the co-owners, spending pre-owned currency from previous auctions with different co-owners would give some participants an unfair advantage. This would result in some users cooperating to share trivial co-owned contents between themselves, and not spend any currencies for the auctions since the contents have no share value at all. This is prevented with group-wise spending, where the currency earned from auctions with some co-owners can only be spend in the future auctions with the same co-owners.

Boundaries: Clarke-Tax mechanism allows users to bid as much as the currency they hold. This free market approach economy adds a level of uncertainty to the auctions, since a participant cannot have a clear opinion about what others might bid. A person that has big earnings can bid with high numbers, and makes little of the taxes, since they can spend more than the others. Limitations to minimum and maximum bids allowed can be beneficial to prevent users that are richer in the currency from dominating the decisions. This also helps agents that participate in the auctions to have better evaluation functions, because they can have a better opinion about the other participants' bids, especially with prior knowledge about the others' characteristics and the context of the content.

With the notion of minimum-maximum boundaries, the balance between currency earnings and expenditures comes into consideration. Users earn currencies by being a co-owner of a content and spend them for decisions in auctions and taxes. Earning currencies equal to or more than the maximum expenditure for an auction would cause the economy to bloat and inflation to emerge. Therefore, earnings from being a co-owner should be less than one can spend for an auction of a content. We propose one to two balance, where the currency earned from a content should be half of the maximum boundary of an auction. Another benefit of this approach is to encourage agents to spend more wisely, since spending less for a relatively unimportant decision could help agents to spend more in the future decisions.

Consider Example 1 where Dave could spend 50 on a picture because he could afford it. With the proposed scheme, group-wise spending would only enable Dave to spend pre-owned currency earned from co-owned contents with the same group. Therefore, the notion of total owned currency becomes obsolete, replaced with group-wise currencies. In this situation, all the co-owners would earn the same amount from the co-owned contents, which would provide equality in auctions. Dave can only bid as much as the other co-owners, so he cannot take advantage by using earned currencies with contents co-owned with other people. The agent-based policy system can also solve the minority dominating the majority problem in Example 2, which is caused by the absence of co-owners in the auction. In this example, Alice and Carol, who do not join the auction which results in the only person, Dave, that doesn't want the photograph to be shared impose his policy on the others. With PANO, every user in the OSN is represented with an agent, which facilitates privacy policies to bid on behalf of their users. Therefore, even if the users are not available for an auction, their agents always are, bidding according to their owners' policies. With the help of automation, Alice and Carol will also be represented in the auction with agents. Since Clarke-Tax mechanism is based on truthfulness, with every co-owner present in an auction, the resulting action will be decided according to everyone's opinion, instead of a minority of the co-owners.

Example 5 Dave wants to share a photograph he is in with Alice, Bob, and Carol. Alice and Carol don't want the picture to be shared, and they want to do a Clarke-Tax auction to decide on the final policy. Dave shared a lot of photographs before, that are unrelated to Alice, Bob, and Carol, and gained bid scores from those contents. Since the mechanism only allows Dave to use currencies obtained from the contents that are related to all Alice, Bob, Carol, and Dave, the scores from unrelated content can't be used to gain advantage. The outcome of the auction is decided over who values their decision most, with the use of only the gained currencies from the previous shared contents of the same group, without any unfair advantage to any participant.

3.3 Bidding Mechanism

Agents bid on auctions based on their privacy policies. As explained in Sect. 3.1, policies have importance values. On top of this, agents also have privacy characteristics, which is the notion of how much privacy-aware an agent is, represented with a value between 0 and 1, named as characteristic coefficient. For a content in an auction, an agent checks its related policies and determines the set of social network users that it wants to share or not share the content with. The characteristic of the agents and the importance of related policies determine how much the agent wants to bid for an auction, according to the given actions of these policies. In Clarke-Tax mechanism, each type of action can receive a different bid from the participants, so even when an agent has conflicting policies (i.e., different policies of an agent preferring different actions for the same network users), it can place bid on conflicting actions, according to their evaluations. Agents should also consider how much currency they own, and place their bids accordingly (e.g., bidding small amounts when short on currency or bidding higher when have enough spendable currency).

In PANO, the bidding mechanism is a linear function that returns an integer value between the bidding boundaries for each action, namely sharing, not sharing and limited sharing, explained in Sect. 2.2. The function considers privacy policies, agent characteristics, the current score balance of the agent and the number of users that are in conflict against other agents in the auction to compute the outcome. For a content, an agent first considers its related policies to find *coefficient* of biddings. This coefficient (*PC*) is calculated as the mean average of the importance values of the policies. To spend the obtained currencies in a prudent manner, we consider both sharing and not sharing actions at the same time for the mean average, if the audience of both policies are the same. That is, we consider not sharing importance as negative, and sharing importance as positive values. For example, if an agent has two policies for a content; one sharing with importance of 0.8 and one not sharing with the importance of 0.2 for not sharing it with the same audience, the importance coefficient is calculated as the mean average of 0.8 and −0.2, thus 0.3 for sharing the content. This coefficient is multiplied by the privacy characteristic coefficient (*CC*) of the agent. The final parameter is calculated according to the bidding boundaries and the current score balance of the agent, namely spendable currency (*SC*). If the agent has two times of the maximum boundary (*mb*) as the balance, the bidding is made with the multiplication of the maximum boundary and the computed coefficient. If the currency is less, the coefficient is multiplied with the agent's owned currency (*oc*) divided into two, in order to save currency for the future auctions. Equation 3 gives the formula generating bids for sharing or not sharing where n is the number of related policies for an auction and $P_\alpha\{i\}$ is the importance value for policy α. Intuitively, an agent generates a high bid if the privacy importance of the image in question is high, the agent values privacy (CC) and it has the resources to spend on the bid (SC).

$$SC = \begin{cases} mb, & \text{if } oc/2 \geq mb \\ oc/2, & \text{otherwise} \end{cases}$$

$$\text{Bid Value} = \sum_{\alpha=1}^{n} (P_\alpha\{i\})/n \times CC \times SC \tag{3}$$

In addition to biddings for sharing or not sharing a content, an agent can also bid for the limited share action. Limited share action aims to share the content with the conflicting set of users, i.e. the users that at least one agent is willing to share the content and at least one agent does not want to share the same content with. For bidding to share with a limited audience, the agent checks how many of the users in the conflicting set of users it wants to share the content with, and simply calculates the ratio of the users fitting into its policies divided by the total number of users in the conflicting users set, namely as fitting ratio (*FR*). The agent bids half of the maximum boundary or half of its own score if it has less, multiplied with the computed ratio for the conflicting users and its characteristic. The formula for computing the bid values for limited sharing is given in Eq. 4, where n is the count of conflicting users considered for the policy, and β is the count of users in the conflicting list fitting into user's own share policies.

$$\text{Bid Value (Limited Share)} = \beta/n \times CC \times SC \tag{4}$$

4 Multi-agent Simulation

In order to evaluate the success of PANO, a multi-agent simulation environment is implemented, where agents represent users in a social network with privacy policies, and use Clarke-Tax auctions to decide on the share policies of co-owned contents. The policies are defined according to rules that regular social network users tend to rely on, and the simulation checks the success of final policy decisions with Clarke-Tax auctions over a chunk of co-owned contents.

4.1 *Context-Based Privacy Constraints*

Social networks users have privacy constraints for the contents they share in the network. These constraints can be represented by rules, which then can be used to model behavior of the users. Kekulluoglu et al. have done a user study to extract privacy rules of frequent users of online social networks [4]. In this work, there are seven decisive rules, which most of them had contextual content properties as a constraint. Two out of these seven rules were eliminated from the current work, since the simulation doesn't support location or event aspects of the social networks. The remaining rules, with relation types and contextual constraints, are presented below:

1. If the user: x is included in a photograph that depicts a depressed mood, don't share the photograph.
2. If the user: x is included in a photograph that is located in a bar then don't share the photograph with family members.
3. If the user: x is included in a photograph that is located on a beach don't share with work related people.
4. If the user: x is included in a photograph that is a mature content, don't share the photograph.
5. Do not share photographs with user: y.

In our policy notation, we define contextual constraints with conditions combined with content types. In the example policy given in Sect. 3, *photograph[scenery]* depicts that the content type is a photograph, and it's categorized as scenery. We represent the photograph categories presented in the rules above with such keywords, which can be obtained with automated photograph tags. In PANO, we assume that the photograph contents are already tagged, so we can assign them into policy conditions. The conditions can have multiple content types, and also multiple constraints for these content types, giving us flexibility in defining the policies. Since the rules in Kekulluoglu et al. [4] focus on not sharing contents in specific conditions, we would require some sharing rules that can create conflicts we aim to resolve in scope of this paper. Therefore, we generate some opposing rules to the ones above, where some agents would like to share the photographs in the same conditions. For example, when an agent has the third rule in their policy, we can generate a policy for another agent about sharing photographs tagged with beach with its friends. In this example, if both agents are co-owners for a photograph on a beach, the intersection of OSN users from work related people for *Agent #1* and friends for *Agent #2* will be in a conflicting situation, because the former do not want to share the photograph while the latter wants to share it.

4.2 Execution of the Simulation

PANO is implemented in Java. The simulation represents social network environments, with the aspects of users, contents, different relation types between users and agents. Agents have defined rules, based on the combinations of relations and contextual tags of photographs, as explained in Sect. 4.1.

As mentioned in Sect. 3, agents are only allowed to use the scores they obtained from earlier shared contents from the same co-owners. Another prevention for abuses is that each agent can only spend a total maximum of 20 scores for each content, so that none of the agents can overvalue it to force its own action. According to the characteristics of the agents, they can disperse their scores to different actions, and they can even decide not to spend any scores if they don't value the sharing or not sharing of the content. After all the bids are taken blindly from each user, the simulation calculates the overall decision and taxes of the agents (if any), and updates their currencies within the network, as explained in Sect. 2.1.

Algorithm 1: Main process of the simulation

```
load the social network;
set relationships;
add contents with contextual tags;
assign co-owners to the contents;
add user characteristics and rules;
foreach Content do
    foreach Co-Owner do
        determine the set of people to share or not share;
        if the decision for a person is differing then the other co-owners then
            remove person from other lists put the person on the conflicting list
        else
            if the person has no share decision & the person not exists in any lists then
                put the person on no share list
            if the person has share decision & the person not exists in any lists then
                put the person on share list
    foreach Co-Owner do
        place bids according to the final lists
    trigger Clarke-Tax mechanism;
    update currencies of the co-owners;
```

Algorithm 1 explains the main process of the simulation. The first five lines initialize the environment by loading the network, setting relationships between the users, adding contents with co-owner assignments and generating agent characteristics and rules. Then the simulation enters into a loop for each content that requires Clarke-Tax auction for deciding the final policy. This loop defines the share, limited share and no share lists of a content between lines 8 and 15. Using the formulas of bid calculations in Sect. 3, agents place bids on each action in lines 16 and 17, and the last two lines update scores according to the outcome of the auction, considering the placed bids and the taxes.

5 Evaluation

The main goal of a privacy management system is to maintain the balance between the system users' privacy requirements and fit the finalized privacy policies as much as it can, even with conflicting policies. We evaluate PANO using the overall system success and user satisfaction metrics presented in Sect. 2.2.

5.1 Correctness of Policies

The first evaluation of PANO is aimed to find out the success of the proposed modifications of the method to the Clarke-Tax mechanism, to prevent abuses and create more satisfactory privacy policies to the community. The success metric defined in Sect. 3 is used to measure the correctness of policies.

Hypothesis 1 Given a set of agents with various characteristics, PANO can assign collaborative policies more successfully than the native Clarke-Tax approach, regardless of the starting point and the level of abuse by the agents, while the higher abuse level decreases the performance of the former method.

Using the success metric, PANO is compared to the pure Clarke-Tax mechanism approach, where group-wise currencies or boundaries are not present. The evaluation is made with five randomly generated social networks of 1000 users, and 200 randomly generated photograph content with contextual tags related to the rules presented in Sect. 4.1. Each network was calculated with considering different starting points where some of the contents are considered shared before. These starting points are decided as 40, 80, 120 and 160 contents shared before, respectively. For representing the pure Clarke-Tax mechanism, two levels of abuse were used in comparisons to model the examples in Sect. 2.2. First level enabled some agents to bid more than the defined 0–20 range, but resulting in consuming all currency if used for every auction. Also, the agents that abuse to protect contents from sharing or try to disseminate to their entire social network are balanced, so the abuser agents can also

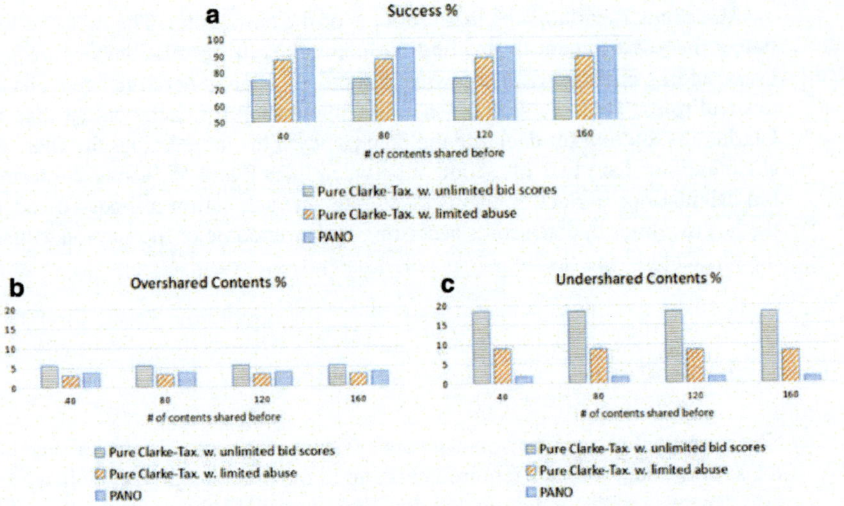

Fig. 1 **a** Percentages of the social network users that are correctly assigned within the final policies according to co-owners' characteristics. **b** Oversharing percentages. **c** Undersharing percentages

have conflicts. Second level of abuse were defined by giving unlimited currencies to the abuser agents, and only letting one-sided abuse for a single auction (i.e. having only agents that want to share the content or doesn't want the content to be shared). After running the simulations with these networks for three different models, the average of satisfactory policy metric for five networks, according to different starting positions is shown in Figure 1a. It shows that with PANO, correct assignments of users with created policies increase, with more than 90% (approximately 95%) success rate. Limited abuse level with pure Clarke-Tax had less than 90% success (approximately 88%), and unlimited currencies for abuser agents gave the lowest results with less than 80% (approximately 76%) success rate.

The same setup is also used for comparing oversharing and undersharing percentages of created policies between the two models. Oversharing covers the users a content is shared with, when some co-owners do not want to share the content with those users. In opposite, undersharing covers the set of users that some co-owners want to share the content with, but it is not shared with this set of users. The results of oversharing and undersharing evaluations are shown in Figure 1b and c, respectively. For oversharing metric, PANO and pure Clarke-Tax method with limited abuse almost performed the same (3.5%), with pure Clarke-Tax approach performing slightly better (less than a half percent). They both performed better than pure Clarke-Tax method with unlimited currency for abuser agents, which had about 5.6% oversharing percentage. The similar performances in oversharing is also investigated, and it has been seen that the abusing agents have more advantage with not sharing the content, since deny-overrides is the default strategy for equal situations. In some cases, limited abuse level had two conflicting abuser agents trying to outbid each other with the agent that prefers not to share always winning when the same bids are made, and oversharing percentages managed to be as low as PANO. For the undersharing metric, this causes both abuse levels with pure Clarke-Tax method to perform far worse than PANO, where PANO had about 1.5% undershare while two levels of abuse methods had about 8.8% and 18.5%, respectively.

In general, PANO performs better than applying pure Clarke-Tax mechanism with both abuse levels, considering all the different starting points. Since the evaluated contents were same for all starting points, the success rates were quite similar in four different setups. Even with the same contents, there is a slight increase in the performance when previously shared content number increases, indicating that with large number of pre-shared contents, Clarke-Tax mechanism will perform better.

5.2 *User Satisfaction in Case of Conflicts*

The second evaluation setup aims to calculate the user satisfaction metric described in Sect. 3 in a larger generated network. The main goal is to create conflicts between users for a large number of contents, and calculate ratios of intended/unintended audience for decided policies of these contents, using PANO and comparing it with the results of the application of a pure Clarke-Tax mechanism.

Hypothesis 2 Given that there is a conflict of policies between agents for every auction, PANO will ensure the user satisfaction of the majority by preventing some agents from abusing the mechanism continuously.

The user satisfaction metric in this evaluation has three measurement types. First is the satisfaction ratio, which shows the percentage of people that the policies of the co-owners matching the final policy, taking the sensitivity of the initial policies of the agents into account. The second and third metrics are the percentages of oversharing and undersharing, which are in the same concept as the first evaluation, but these also include the sensitivity levels of the agent policies.

The setup of the second evaluation is constructed with a 1000 agent network, where 100 agents owned contents and had approx. 250 friends each. 500 contents were generated with tags, and randomly assigned to 4 of the 100 agents within the network. Policies of the agents were also randomly generated with different sensitivity levels. For each content, at least one conflicting policy were assured with additional policy generation for some of the co-owners, to create 1 against 2, 1 against 3, or 2 against 2 conflicts. Clarke-Tax mechanism is run with PANO and the pure approach, where at least one agent was trying to abuse the mechanism to get decisions for sharing or not sharing the content according to its own policy. The user satisfaction, oversharing, and undersharing metrics were calculated at each 100 content mark according to equation (1), and these three comparisons are presented in Fig. 2.

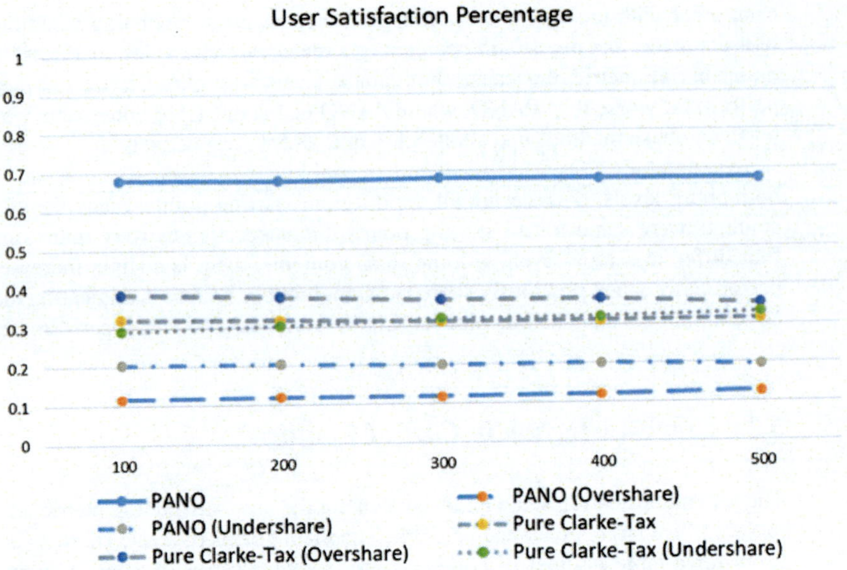

Fig. 2 Comparison of user satisfaction

6 Discussion

Applying multiple privacy preferences from several users for a single OSN content to define a single privacy policy is a major challenge, and it can even be more complicated when these preferences are conflicting. Several studies worked on collaborative privacy management, applying different methods with different approaches. Role-based access control, which was applied widely to other software or operating systems, is not fully suitable to OSNs, since it can't capture binary or context-dependent relationship features that are available in most OSNs. Fong [3] introduces Relationship Based Access Control (ReBAC) mechanism and provides a model to make it applicable to OSNs, where users can define their privacy constraints related to the relations that are available in OSNs, such as friends or colleagues. ReBAC is one of the earliest works which included user-related privacy constraints in access control decisions in OSNs, and provided a basis for many collaborative privacy management mechanisms.

Such and Rovatsos [11] propose a model, where predefined policies of the users are used to find out conflicts for contents and a middle ground is found with a negotiation mechanism. Such and Criado [10] extend that work by modeling user behavior and using a software mediator where some negotiation actions are made without direct user input. Kekulluoglu et al. [4] use multi-agent negotiation as well but apply a more comprehensive negotiation protocol. They further take into account incentives for agents. Kokciyan et al. [5] use argumentation to resolve privacy disputes. In that

Table 3 Comparison of desirable properties

Properties	Negotiation	Argumentation	Pure Clarke-Tax	PANO
Automation	✓	✓	✗	✓
Fairness	✗	✓	✗	✓
Concealment	✗	✗	✓	✓
Protection before exposure	✓	✓	✓	✓

work, OSN users were represented by software agents, where the agents have access to domain knowledge and infer semantic rules. With argumentation mechanism, agents can attack other agents' assumptions to convince the others to accept its users' privacy requirements. While this is successful, it requires agents to be able to reason semantically, which may not be possible in various environments.

Mester et al. [7] propose four desirable properties for a privacy management system, namely automation, fairness, concealment of privacy concerns, and protection before exposure. Automation property is the capability of the system to perform without human interference. Fairness property depicts the system's ability to provide equality to its members. Concealment of privacy concerns is about the coverage of the members' privacy policies; the more agents can hide their policies from the others, the more the property is satisfied. Protection before exposure is related to the privacy management method's application before the related entity is published in the system, causing possible violations. We compare our method with some other approaches for privacy management in terms of these properties in Table 3. The compared approaches are negotiation, argumentation, and pure Clarke-Tax approach with no modifications.

According to the comparison in Table 3, negotiation approaches for privacy management can be designed to be fully automated, even though some methods require human intervention. It can cover fairness to some extent, but it is still open to abuses by the negotiating agents. Since agents have to expose their privacy requirements for negotiation, they cannot offer the concealment of privacy concerns property. Negotiation protects the agents before exposure, but it might require multiple iterations over an algorithm to finalize a privacy policy. Argumentation is similar with negotiation for four of the properties, but it also provides fairness, since the agents are given the opportunity to attack others' arguments over multiple iterations to decide the final policy, thus defending themselves against unfair arguments. Pure Clarke-Tax mechanism provides a one-shot, simple algorithm, concealing privacy concerns of the agents, while being easy to compute. However, it is used by social network users, therefore with no automation, and is open to several types of abuses, as mentioned in Sect. 2.2. PANO is fully automated, where each user is represented with an agent, and agents act on behalf of their owners according to their privacy policies. Measures to prevent abuses to present a fair mechanism is explained in Sect. 3, which satisfies the desirable fairness property for such systems. prevent the system from having the first and the second properties.

Future work of the methods presented in this paper will focus on modeling user behavior and predicting the decisions of users with software agents. We plan to achieve this with a user study where participants can act according to their privacy concerns for co-owned contents and6 bidding with the same mechanism of PANO. With the results of the user study, we will cluster users according to their privacy concerns and behavior in OSNs. Using these clusters with a semantically related, hierarchical content categorization will be a guidance to implement efficient software agents, which can be used to represent OSN users more precisely with less amount of prior knowledge within our Clarke-Tax based Collaborative Privacy Management model.

References

1. E. Clarke, Multipart pricing of public goods. Publ. Choice **11**(1), 17–33 (1971)
2. E. Ephrati, J.S. Rosenschein, The Clarke tax as a consensus mechanism among automated agents, in *Proceedings of the Ninth National Conference on Artificial Intelligence, AAAI '91*, vol. 1 (AAAI Press, 1991), pp. 173–178
3. P.W.L. Fong, Relationship-based access control: Protection model and policy language, in *Proceedings of the First ACM Conference on Data and Application Security and Privacy, CODASPY '11* (ACM, New York, NY, USA, 2011), pp. 191–202
4. D. Kekulluoglu, N. Kokciyan, P. Yolum, Preserving privacy as social responsibility in online social networks. ACM Trans. Internet Technol. **18**(4), 42:1–42:22 (2018)
5. N. Kökciyan, N. Yaglikci, P. Yolum, An argumentation approach for resolving privacy disputes in online social networks. ACM Trans. Internet Technol. **17**(3), 27:1–27:22 (2017)
6. A. Lampinen, V. Lehtinen, A. Lehmuskallio, S. Tamminen, We're in it together: Interpersonal management of disclosure in social network services, in Proceedings of the SIGCHI Conference on Human Factors in Computing Systems, CHI '11 (ACM, New York, NY, USA, 2011), pp. 3217–3226
7. Y. Mester, N. Kökciyan, P. Yolum, *Negotiating Privacy Constraints in Online Social Networks* (Springer International Publishing, Cham, 2015), pp. 112–129
8. A.C. Squicciarini, D. Lin, S. Sundareswaran, J. Wede, Privacy policy inference of user-uploaded images on content sharing sites. IEEE Trans. Knowl. Data Eng. **27**(1), 193–206 (2015). Jan
9. A.C. Squicciarini, M. Shehab, F. Paci, Collective privacy management in social networks, in *Proceedings of the 18th International Conference on World Wide Web, WWW '09* (ACM, New York, NY, USA, 2009), pp. 521–530
10. J.M. Such, Natalia Criado, Resolving multi-party privacy conflicts in social media. IEEE Trans. Knowl. Data Eng. **28**(7), 1851–1863 (2016). July
11. J.M. Such, M. Rovatsos, Privacy policy negotiation in social media. ACM Trans. Auton. Adapt. Syst. **11**(1), 4:1–4:29 (2016)

Prosocial or Selfish? Agents with Different Behaviors for Contract Negotiation Using Reinforcement Learning

Vishal Sunder, Lovekesh Vig, Arnab Chatterjee, and Gautam Shroff

Abstract We present an effective technique for training deep learning agents capable of negotiating on a set of clauses in a contract agreement using a simple communication protocol. We use Multi-Agent Reinforcement Learning to train both agents simultaneously as they negotiate with each other in the training environment. We also model selfish and prosocial behavior to varying degrees in these agents. Empirical evidence is provided showing consistency in agent behaviors. We further train a meta-agent with a mixture of behaviors by learning an ensemble of different models using reinforcement learning. Finally, to ascertain the deployability of the negotiating agents, we conducted experiments pitting the trained agents against human players. Results demonstrate that the agents are able to hold their own against human players, often emerging as winners in the negotiation. Our experiments demonstrate that the meta-agent is able to reasonably emulate human behavior.

1 Introduction

Reaching a consensus on the contents of a contract agreement is often an expensive and time-consuming task for the negotiating parties involved. Automating this process is important not just from an industrial point of view but also for all related problems where negotiation is required. The primary challenge in this endeavor is to suitably model the behavior of an *AI agent*. Several papers in the literature

V. Sunder (✉) · L. Vig · A. Chatterjee · G. Shroff
TCS Research, New Delhi, India
e-mail: s.vishal3@tcs.com

L. Vig
e-mail: lovekesh.vig@tcs.com

A. Chatterjee
e-mail: arnab.chatterjee4@tcs.com

G. Shroff
e-mail: gautam.shroff@tcs.com

T. Ito et al. (eds.), *Advances in Automated Negotiations*, Studies in Computational Intelligence 905, https://doi.org/10.1007/978-981-15-5869-6_5

[4, 14] have addressed the problem of modeling agents for contract negotiation. However, most of the proposed approaches use a combination of mathematical models and rules. With the upsurge of *deep reinforcement learning*, in the past few years, it becomes interesting to analyze the behavior learnt by *deep* agents trained to negotiate via reinforcement learning. Recently, we have seen valuable contributions in the domain of *AI agents'* negotiation using deep reinforcement learning [3, 17] but much work needs to be done for their direct applicability to various negotiation-oriented domains such as contract negotiation.

Although there is training data available for a negotiation domain in which agents learn to divide items between themselves [17] (this is the problem of resource reallocation which has been the formalism in most negotiation studies), no such data is available for contract negotiation where agents need to reach a consensus about which items to share between them (clauses). This makes it difficult to train an agent which can converse in natural language. Although crowdsourcing can be utilized for generating data,it is resource intensive. Nevertheless, we present an approach where we train agents to converse with each other using a simple protocol which is an interpretable sequence of bits. This protocol is scalable and robust. The training is done using reinforcement learning. We first model an agent as a neural network and then train two such agents concurrently where they play several rounds of negotiation games against each other and learn to coordinate with each other based on the outcome (reward). The behavior of the agents is modeled by using the effective technique of varying the reward signal. In this way, we get agents with different behaviors. We show that agents trained in this manner indeed learn to coordinate their moves and produce context-relevant outputs. We observe that agents have consistency in their behaviors and this is shown by making all agents negotiate against each other and analyzing the results.

For emulating human behavior, we train an agent (*meta agent*) with a dynamic behavior. We use a model which is an ensemble of the 4 agents mentioned above, with a "selector" agent on top which selects the appropriate behavior based on the negotiation state. This agent is also trained using reinforcement learning. We have found evidence that the meta-agent learns a simple decision tree while selecting its behavior.

Finally, for assessing the usability of the negotiation agents in real-world scenarios, we conduct experiments where we evaluate the agents against human players. We show that agents show consistency in behaviors even against human players. Thus, we show that our agents are deployable in real industrial scenarios for negotiating on a contract.

Hence, the main contributions of this paper are:

1. A deep learning model and a reinforcement learning procedure for training an AI agent to negotiate in the domain of contract negotiation.
2. Modeling selfish/prosocial behavior by varying the reward signal for the agent and its opponent in the reinforcement learning framework and empirical evidence for the same.

3. An AI agent with a dynamic behavior (varying within a negotiation instance) by learning an ensemble of different agent behaviors using reinforcement learning.
4. Evidence for the usability and success of the negotiation agent against human players through real-life experimental results.

The rest of the paper is organized as follows: Sect. 3 formulates the problem statement and provides a general framework for our negotiation environment. It also describes the model for the negotiation agent and the evaluation metrics. Section 4 describes the model architecture and learning procedure. Section 5 analyzes the results when the agents which are trained against each other negotiate. Section 6 provides an analysis of the behavior of the agents and demonstrates consistency in the learnt behaviors. Section 7 provides details about the model and performance of the *meta agent* which has a mixed/dynamic behavior. Section 8 gives the results of the human evaluation. Finally, we conclude and give directions for future work in Sect. 9.

2 Related Work

Contract negotiation can be seen formally as a resource reallocation problem [6, 7]. Several papers have addressed the problem of building AI agents that can negotiate on a contract. But most have focused on domain-specific applications of the same [2, 26]. However, our approach to this is one where we require agents to reason about a shared set of decision variables (clauses) [9]. The problem formulation is similar to the furniture layout task [8, 25], where two conversants try to agree on five pieces of furniture to place in a room.

In general, much work on building negotiation agents has focused on mathematically modeling these agents or building fine-grained rules [2, 4, 23, 27]. However, machine learning has seldom been used for tackling this problem. One of the initial works on the use of reinforcement learning in negotiation focused on formally defining a reinforcement learning state for the task [13].

Policy Hill Climbing (PHC) has been shown to converge better than traditional algorithms such as Q-Learning [11] for multi-agent training, but convergence becomes an issue with an increase in size of the *action space*. This is where modern *Deep Learning* methods become useful.

With recent advancements in deep learning, there has been a lot of interest in building agents that learn to communicate and coordinate for various multi-agent tasks. Deep reinforcement learning methods can be used to train agents to cooperate in complex environments [16]. Furthermore, it is possible to build deep learning agents that can be trained to develop their own protocols to solve collaborative tasks [3, 10].

More recently, we see valuable contributions in building *deep* agents for negotiation domain. With the availability of data, deep learning agents can be trained to imitate humans (converse in natural language) and subsequently even outperform them by using reinforcement learning techniques [17]. Although robust and end to end, such models require training data, collecting which can be resource intensive for different domains.

Modeling negotiation behavior by varying the reward signal has been used previously in Ref. [3], but their main aim is to study the communication protocols that can emerge during negotiation. They limit themselves to two behaviors whereas we explore four behaviors in addition to a dynamic behavior agent.

With the exception of Ref. [17], none of the previous works, to the best of our knowledge provide qualitative/quantitative results against human players.

3 Problem Formulation

Here, we formulate the problem of negotiating on a set of clauses in a contract agreement scenario wherein two parties have to reach a consensus as to which clauses need to be included in/excluded from the final draft. For this paper, we have considered 6 clauses but our method is generalizable for any number of clauses (We have considered 6 clauses because we found that this reduces the training time of our model while preserving enough complexity in the problem).

3.1 Negotiation Environment

The value that an agent attaches to the clauses is represented by a utility function which is a vector of 6 integers between −12 and 12 (excluding 0) such that their sum is zero. There is an additional constraint that there is at least one positive and one negative value in this vector and that the sum of positives is +12 and that of the negatives is −12. More formally, this vector is represented as:

$$U = \text{Shuffle}(P \bigoplus N) \tag{1}$$

$$P = [p_1, p_2, \ldots, p_k] \tag{2}$$

$$N = [n_1, n_2, \ldots, n_{6-k}] \tag{3}$$

where, $0 < k < 6$, $\bigoplus$ is the concatenation operator and Shuffle(.) is a "random shuffling" function. Also, $p_i \in \{1, \ldots, 12\}$ and $n_i \in \{-12, \ldots, -1\}$, along with the following constraints:

$$\sum_i p_i = 12 \tag{4}$$

$$\sum_i n_i = -12 \tag{5}$$

Each element in the list represents the "importance" that the agent attaches to the corresponding clause. We have this distribution so that in every case there is a mixture of the most beneficial clauses (values summing to 12) and the most harmful clauses (values summing to -12). For example, if an agent has a utility $[9, -5, 2, -1, -6, 1]$, clause one is non-negotiable as it has the highest value. On the other hand, clauses three, four and six might be negotiable as they do not have a high absolute value (Of course, the decision as to whether a clause is negotiable or not has to be made the agent). Notice that the maximum point an agent can score is 12 and the minimum point is -12. We exclude the "zero-valued" (neutral) clause to avoid skewness of values in the utility function.

Each of the agents receives this utility which is sampled uniformly. The agents then communicate with each other by giving offers, which is a sequence of 6 bits, $S_t \in \{0, 1\}^6$. Here, subscript t refers to the time step at which the offer was produced. Each bit in this sequence is the agent's decision on the corresponding clause (0 meaning exclude and 1 meaning include). The communication follows a sequential structure and the agent that goes first is decided by an unbiased coin flip. This communication goes on between agents until one of the following happens:

1. An agreement is reached. This happens when an agent gives the same offer that it receives.
2. The time runs out. We keep a limit of 30 offers (15 for each agent) after which the negotiation process stops with a disagreement.[1]

At the end of the negotiation, each of the negotiating parties gets a reward based on the agreed sequence of bits. So, if agents A and B have utilities U^A and U^B respectively, and the agreed sequence is S, A gets $S.U^A$ and B gets $S.U^B$, where . represents the dot product.

An example negotiation between two agents is illustrated in Fig. 1.

3.2 *Negotiation Agent*

Given a utility function and the opponent's offer, we divide the problem of producing an offer into two parts:

1. *Deep Neural Network (DNN) Component*: The first part is to decide the number of bits to be flipped in the opponent's offer. This part is taken care of by the deep neural network which is trained through *Reinforcement Learning*.

[1] We have run our experiments for different number of clauses and thresholds and have found that the agents require a certain minimum number of turns during training so that they learn to reach an optimal solution. For 6 clauses, this turns out to be 30. A more thorough analysis of the relation between number of clauses and the threshold may form an interesting subject for future work.

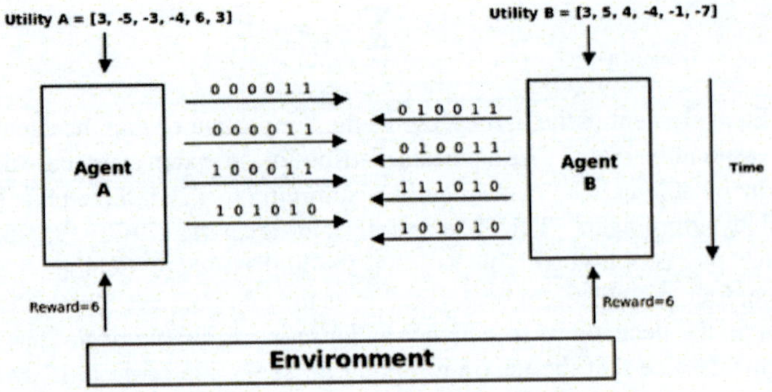

Fig. 1 A negotiation example

2. *Rule-Based Component*: Once the neural network decides the number of bits to be flipped, the part of deciding which exact bits to flip is deterministic, i.e., flip the bits that result in maximum increase in score.
   For example, if the utility is $[2, -6, -2, -4, 7, 3]$, the opponent's offer is $[1, 1, 1, 0, 0, 1]$ and the number of bits to be flipped is 3 (decided by the neural network), we would flip the second, third and fifth bit (rule based) as that would increase the score by $6 + 2 + 7 = 15$ points and we would get 12 points (which is the maximum one can get).

   This approach has two distinct advantages:

1. The action space for reinforcement learning reduces significantly. This is particularly advantageous when the number of clauses is much higher than 6. In general, for n clauses, the action space goes from 2^n to n and hence simplifies the training procedure.
2. The rule-based component provides us leverage over the agent making it suitable for deployment against humans (see Sect. 8). Without the rule-based component, the deep network might learn policies which are suitable for negotiation with AI agents but not against humans [3, 17].

We call this negotiation agent the *RL Agent*.

3.3 Evaluation Metrics

The evaluation of a negotiation is done based on the following metrics:

3.3.1 Dialog Length

The dialog length is the average number of time steps for which the negotiation lasts.

3.3.2 Agreement Rate

Percentage of negotiations that end in an agreement represents the agreement rate.

3.3.3 Optimality Rate

The optimality rate is the percentage of negotiations that end in an optimal deal. We say that a deal is optimal if it is both *Pareto Optimal* [22] and the both agents receive a positive score. A solution is Pareto optimal if neither agent's score can be improved without lowering the other's score.

Average score

The average number of points earned by an agent. To see the maximum joint reward the agents can earn on an average on optimal deals, we greedily search through all possible deals ($2^6 = 64$) for all of the samples in the test set and select the one which results in the maximum joint reward and optimal deals. The average of maximum joint reward for the test set is 1.40 (0.70 for each agent.[2] This is used for comparison of average scores of trained agents in Tables 2 and 3).

4 Learning Procedure

We define below, the architecture and training procedure for the DNN component of the negotiation agent.

4.1 Model Architecture and Learning

At a given time step t in the negotiation process, an agent A receives the following as the *state* input:

1. It's Utility function U^A.
2. Offer given by opponent B, S_t^B.
3. It's previous offer, S_{t-1}^A.
4. Agent ID, $I \in \{0, 1\}$.

We convert this input into a dense representation D_t^A as

[2]We normalize the points earned by the maximum points possible, i.e., 12.0.

$$D_t^A = [\text{OfferMLP}([U^A, S_t^B]), \text{OfferMLP}([U^A, S_{t-1}^A]),$$
$$\text{AgentLookup}(I), \text{TurnLookup}(t)]. \tag{6}$$

Here OfferMLP(.) is a 2-layer MLP with ReLU activation [19], AgentLookup(.) is an embedding which gives a dense representation for the agent identity and TurnLookup(.) is another embedding which encodes information in the time step.

The representation D_t^A is passed to a 2-layer GRU [5] as

$$h_t^A = \text{GRU}(D_t^A, h_{t-1}^A), \tag{7}$$

where h_{t-1}^A is the hidden state generated by A at its previous turn. The number of bits to be flipped is predicted (action taken) by sampling from policy π_A:

$$\pi_A = \text{Softmax}(Wh_t^A). \tag{8}$$

During test time though, we do not sample and select the action with the highest probability.

At the next time step, the agent B also outputs a similar policy π_B. Each agent $i \in \{A, B\}$ tries to maximize the following objective individually:

$$L_i = \mathop{\mathbb{E}}_{x_t \sim (\pi_A, \pi_B)} \left[\sum_t \gamma^{(T-t)} (r_i(x_{1...T}) - b_i) \right] + \lambda H[\pi_i]. \tag{9}$$

Here,

1. x_t is the action taken by an agent at time t,
2. γ is the *discount factor*,
3. T is the total time steps for which the negotiation lasts,
4. $r_i(x_{1...T})$ is the reward received by agent i at the end of the negotiation which is a function of the sequence of actions x_t taken by the agent from $t = 1$ to $t = T$,
5. b_i is the baseline which is used to reduce variance, and
6. $H[\pi_i]$ is the entropy regularization term to ensure exploration and λ controls this degree of exploration [18].

The gradient of L_i is estimated and the parameters updated as in REINFORCE [24].

The parameters of agents A and B are shared. These parameters are updated after each episode. An episode refers to a negotiation game between agents A and B. We run the training for 5 epochs with 10^5 episodes in each epoch. The values of all hyperparameters are given in Appendix 1.

Table 1 Behavior induced in agents A and B depending on whether there is an optimality signal in the rewards. The agents in the first two rows are subject to the same signal. Hence, they have similar tags in the brackets

Behavior		Signal	
A	B	A	B
Prosocial (Agent 1)	Prosocial (Agent 1)	Yes	Yes
Selfish (Agent 2)	Selfish (Agent 2)	No	No
Prosocial (Agent 3)	Selfish (Agent 4)	Yes	No

4.2 Behavior Modeling (Reward)

The manner in which rewards are given to the RL agent decides its behavior. In particular, we enforce *selfish* and *prosocial* behavior in the RL agent in the following way (although there may be a plethora of complex behaviors, we limit ourselves to the simplest ones):

1. To enforce prosocial behavior, the agent is given the reward (the number of points earned at the end of the negotiation) iff the deal is *optimal*. If the deal is not optimal, the agent is given a reward of −0.5. This ensures that the agent not only cares about its own gain/loss while learning its policy but also takes into account the opponent's priorities as well. In other words, the reward here has a signal for the overall optimality.
2. If there is no optimality signal in the reward, i.e., the agent receives as a reward, whatever it earned in the negotiation, then a selfish behavior is induced. The agent then learns to maximize its own score.

Both agents receive a reward of −0.5 if the negotiation ends in a disagreement.

As two agents learn concurrently, we get agents with four different behaviors depending on how the opponent is trained to behave (Table 1).

1. *Prosocial agent trained against a prosocial agent (PP)*: We get the behavior *PP* when both agents are trained to have a prosocial behavior.
2. *Selfish agent trained against a Selfish agent (SS)*: If both agents are trained to be selfish, we get the agent *SS*.
3. *Selfish agent trained against a Prosocial agent and vice-versa (SP, PS)*: When one agent is trained to be selfish and its opponent is trained to be prosocial, we get two agents whom we call *SP* and *PS* respectively.

5 Coordination Between Trained Agents

Here we analyze the results of the negotiation between agents who have been trained against each other. To assess whether these agents are learning to coordinate in a way so as to improve their scores while maintaining their enforced behavior, we

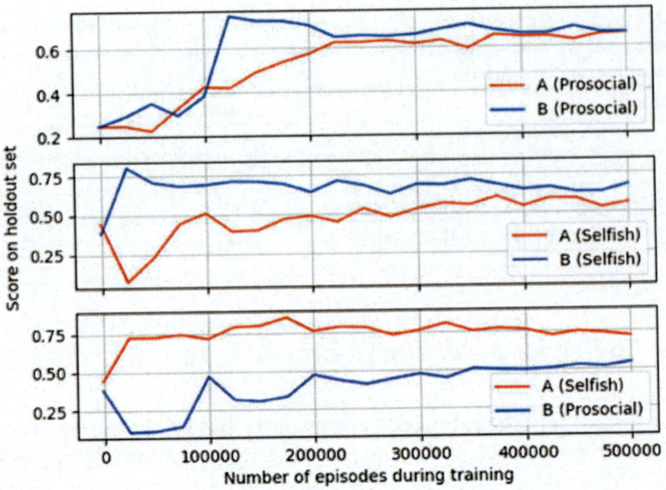

Fig. 2 Learning curves

make the agents play a set of 2000 negotiation games which are held out, after every 25,000 episodes of training. Figure 2 shows this performance. From these curves, we observe the following:

1. When both agents are trained to be prosocial, they gradually converge to the same score. This is not surprising as both agents are being trained to reach a *middle ground*.
2. When both agents are trained to be selfish, one of the agents ends up scoring more than the other. This, we observe is because it is impossible to reach an agreement (which is the ultimate goal) if both agents are stubborn hence it is necessary for at least one agent to *compromise*.
3. When one agent is trained to be selfish and the other to be prosocial, the selfish one clearly outscores the prosocial one because the selfish agent and will seldom concede. The prosocial agent on the other hand, concedes more easily.

To test whether the RL agent learns something non-trivial, we compare its performance with two simple baselines:

1. *RANDOM* At every step, the agent chooses random number of bits to be flipped.
2. *COMMON* Agent 1 (agent who goes first) gives its most selfish offer followed by Agent 2 who does the same. At the third step, Agent 1 offers the intersection of the first two offers (common benefit) to which Agent 2 agrees. If there is no intersection, it is a disagreement.

The results in Table 2 are an average over a separate test set of 30,000 negotiations. It is clear from the results listed out in the table that all three variants of behavioral combinations do better than the baselines in terms of Optimality and Joint Reward.

Table 2 Coordination between trained agents. In optimality column, the numbers in bracket are the percentages on the agreed deals

Agent A	Agent B	Dialog length	Agreement rate (%)	Optimality rate (%)	Average score	
					A(0.70)	B(0.70)
RANDOM		15.90	100	24.55	0.25	0.25
COMMON		3.77	79.54	70.39 (88.49)	0.50	0.50
PP	PP	16.98	96.24	82.33 (85.55)	0.65	0.66
SS	SS	17.47	88.31	74.88 (84.79)	0.54	0.69
SP	PS	13.87	91.90	86.74 (94.38)	0.73	0.55

This shows that the agents which are trained against each other learn to *coordinate* their moves such that apart from maintaining their enforced behavior, they try to maximize their scores as well as the optimality. Observe that joint reward is maximum when both agents are prosocial as both agents are not only concerned with maximizing their own reward but also of their opponent's so as to reach optimal deals. This corroborates the results reported in Ref. [3].

Figure 3 shows the frequency distribution of the sequence of actions (an action is the number of bits flipped by an agent in the opponent's offer) taken by the RL agents during the course of negotiations during test time. In each bar, the orange part

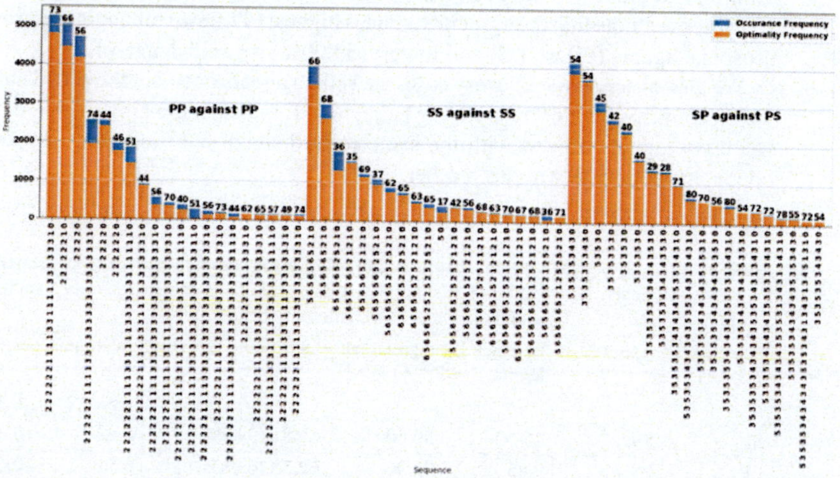

Fig. 3 Frequency distribution of the sequence of actions taken by the agents for all three cases. This only shows the top 20 occurrences. The orange part represents the number of optimal deals when the *agents decide* to use the corresponding sequence. Each point on the x-axis represents the sequence of number of bits flipped by the agents one after the other. The number on the top of each bar is the percentage optimality using the given sequence throughout

represents the number of optimal deals. The distributions show that there is a joint preference among agents for certain sequences more than others which is evident by their *skewed* nature. This nature of the distributions also raises the question: Do the agents learn to identify a context from their utilities and act according to that or do they learn to give generic outputs irrespective of their utilities? To answer this question, we made the agents negotiate on the test set using only the 20 most frequent sequences. The number on the top of each bar is the percentage optimality using the given sequence throughout the test set. We observe that none of the numbers are greater than the overall optimality reported in Table 2. This shows that the agents indeed capture the context from their utilities and behave accordingly.

6 Interplay Between Agents

In the previous section, we analyzed the results when an agent negotiates with the agent against which it is trained. But to analyze the performance of an agent against a policy it has never seen, we run the test negotiations between agents who have never seen each other during training. These negotiations are what we refer to as *Interplay Negotiations*. Table 3 shows the results of these negotiations. These results are an average over the test set of 30,000 negotiations.

It is clear from the results that the optimalities of the interplay between agents are not very high which is because these agents have never seen each other during training and thus have not been able to develop their policies accordingly.

Moreover, note that the agreement rate is highest (97.96%) for negotiation between prosocial agents (PP vs. PS) and lowest (59.00%) for selfish agents.

We also observe that in every case, the selfish agents outscore the prosocial agents which confirms their corresponding behaviors. It is also interesting to note the scores when two agents trained with the same reward signal but trained against different opponents negotiate with each other.

Table 3 Interplay between Agents. In optimality column, the numbers in bracket are the percentages on the agreed deals. As we get two SSs and PPs after training, we choose the ones which outscore its opponent during training and use them in the interplay negotiations

Agent A	Agent B	Dialog length	Agreement rate (%)	Optimality rate (%)	Average score	
					A(0.70)	B(0.70)
SP	SS	26.50	59.00	55.81 (94.59)	0.42	0.48
PP	PS	9.85	97.96	62.55 (63.85)	0.51	0.68
PP	SS	23.98	90.01	69.80 (77.54)	0.44	0.75
SP	PP	24.64	90.43	64.28 (71.08)	0.71	0.45
SS	PS	11.89	93.03	69.43 (74.63)	0.70	0.50

Table 4 Difference in scores for all agent combinations. The selfishness of an agent decreases from left to right and top to bottom

		Player B			
Player A		SS	SP	PS	PP
	SS	–	0.06	0.20	0.31
	SP	–	–	0.18	0.26
	PS	–	–	–	0.17
	PP	–	–	–	–

1. When a selfish agent trained against a selfish agent (SS) negotiates with a selfish agent trained against a prosocial agent (SP), the former outscores the latter by a margin of 0.06 points. This establishes the fact that SS is more selfish than SP.
2. When a prosocial agent trained against a prosocial agent (PP) negotiates with a prosocial agent trained against a selfish agent (PS), the latter outscores the former by a margin of 0.17 points. This shows that PP has learnt to be more generous than PS.

From this, and from other interplay negotiations in Table 3 we observe varying degrees of selfish/prosocial behavior in agents with some agents being more selfish than others.

To verify the consistency in agent behavior, we have shown the differences in scores (Player A–Player B) for all interplay negotiations in the form of a matrix in Table 4. Here, each entry is the difference in scores when corresponding agents negotiate. We observe that the differences are in increasing order along every row and decreasing along columns. As the agents are arranged in decreasing order of their selfishness (from left to right and top to bottom), this kind of distribution confirms consistency in their behavior (i.e., if A beats B with a margin m and B beats C, then A should be able to beat C with a margin greater than m).

7 Meta-Agent—Modeling Dynamic Behavior

On the surface, the selfish agent seems to be best as it always outscores its opponents. But using such an agent leads to many disagreements if the opponent is also selfish (see first row in Table 3). This has also been observed in [12] apart from the fact that selfish and prosocial behavior are not, in general, separable processes in negotiation. It is also important to note that humans don't really negotiate using a fixed policy (either prosocial or selfish). They tend to follow a mixed behavior with some degrees of both depending on the state of the negotiation process. According to previous work [1], there is no universally best policy to negotiate and that it depends on the nature of the opponent. With this motivation, we try to model one policy that works well against all agents by using a mixture of agents. We do this by training another RL agent to choose which of the four agents' policies to use given the state of the negotiation.

7.1 Model and Training

The meta-agent is an ensemble of all four agents, with a *Selector Agent* on top. The job of the selector agent is to select the output offer of one of the four agents given the context U. This agent is modeled as a neural network similar to the one described in Sect. 4.1, except that it also takes the output of all four agents as part of its state input. The selector agent outputs a policy π_s from which an action is sampled. This action is the offer produced by one of the four agents. The selector agent maximizes the following objective:

$$L_s = \mathop{\mathbb{E}}_{s_t \sim \pi_s}\left[\sum_t \gamma^{(T-t)}((r_s(s_{1...T}) + r_o) - b_s)\right] + \lambda H[\pi_s], \tag{10}$$

where $r_s(s_{1...T})$ is the reward that meta-agent gets at the end of the negotiation which is a function of the sequence of actions s_t it takes and r_o is the reward of the opponent. Note that we are giving the joint reward to the meta-agent which is a simple way of ensuring that it is not biased toward one particular agent while selecting.

For training, we randomly select one of the four agents as the opponent and make it play a batch of 100 negotiation episodes with the meta-agent. During this process, we freeze the weights of the opponent. Similar to Sect. 4.1, we run 10^5 episodes for 5 epochs.

7.2 Analysis

We make the meta-agent negotiate against each of the four agents one by one on the test set and the results are reported in Table 5. In terms of the scores, the meta-agent is able to outscore the prosocial agents but not the selfish ones. The meta-agent does well to coordinate with all agents with is reflected by the optimality. Also, the joint reward for all the cases is greater than 1.20. In spite of that, it is not able to match on results reported in Table 2.

Selector learns a decision tree: To analyze what the selector agent actually learns to do, we look at the frequency distribution of the agent selection sequence that it follows against all four agents. This is shown in Fig. 5. The interesting part of all the four distributions is the x-axis which is the sequence of agent selection. It is same for all four cases. Moreover, every sequence is a subsequence of some larger sequence. This suggests that the selector agent learns to follow a decision tree as shown in

Table 5 Performance of meta agent against all four kinds of agents

B	Dialog length	Agreement rate (%)	Optimality rate (%)	Average score	
				Meta	B
PP	18.68	94.41	77.15 (81.71)	0.64	0.61
SS	19.17	86.25	73.33 (85.02)	0.54	0.66
PS	13.10	92.27	76.56 (82.97)	0.69	0.55
SP	20.53	90.22	81.40 (90.22)	0.55	0.71

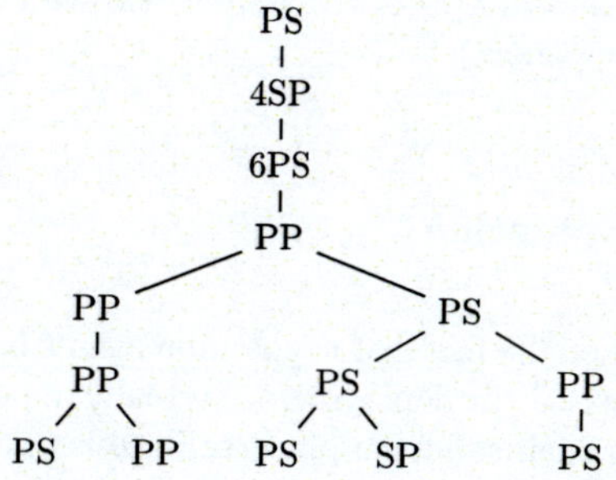

Fig. 4 Decision tree learnt by selector agent. The numbers 4 and 6 at the second and third node denote a repetition of SP and PS 4 and 6 times respectively

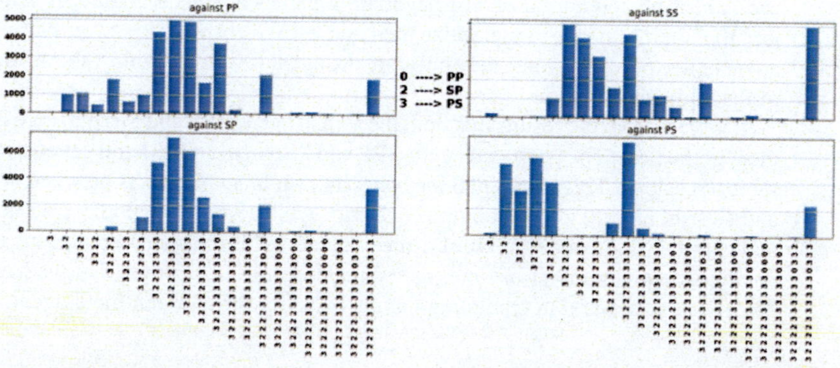

Fig. 5 Frequency distribution of agent selection by the meta-agent. Each point on the x-axis represents the agent selection sequence by the meta-agent. Each number in this sequence corresponds to a particular agent

Fig. 4. Here, every node is the agent selected by the selector at a given state and the edge represents the transition from one state to the next.

The fact that the selector agent learns a decision tree suggests the following:

1. The agent learns just one policy (the simplest) which works against all agents.
2. We know that it is difficult for an agent to decipher the behavior of the opponent until after a few moves, hence it makes sense to learn just one policy which works well at any stage.

The selector agent starts off with being selfish (SP) and then becomes more generous (PS and PP). This seems like a reasonable policy where an agent tries to maximize its score initially by making selfish offers but starts giving prosocial offers toward the end so as to reach a deal.

Moreover, observe that the selector agent never selects SS as that is the most selfish agent and the chances of reaching a deal with this agent are less (especially against a selfish opponent).

8 Human Evaluation

We have established the fact that negotiation agents have learnt to negotiate against each other. But for real-life deployment, it is really important to test their performance when negotiating against human players. For this purpose, we organized an experiment in our lab where our colleagues played several rounds of negotiation games with all the five negotiation agents (PP, SS, SP, PS and META). A total of 38 human players negotiated for 3 rounds of negotiation against all 5 agents. This means that each agent played a total of 114 negotiation games against humans. Humans were told that their aim was to maximize their scores. We ensured this by providing them an incentive for every game they outscore the agent (reward). Table 6 shows the result of the human evaluation.

It is clear from the results that both the selfish agents (SS and SP) outscore humans most of the time. The prosocial agents (PP and PS) on the other hand get outscored on more occasions. This shows that the behavior of human players is between prosocial and selfish, i.e., they have a hybrid behavior which we had hypothesized in Sect. 7 and that had motivated us build a meta-agent. With the meta-agent, humans win an almost equal number of times as the meta-agent. This proves that we have been somewhat successful in emulating human behavior through our meta-agent.

Table 6 Results of human evaluation

Agent	Dialog length	Agreement rate (%)	Optimality rate (%)	Agent score	Human score	Agent won (%)	Human won (%)	Tied (%)
PP	15.07	87.38	70.87	0.58	0.62	36.67	51.11	12.22
SS	19.56	73.79	60.20	0.58	0.44	60.53	21.05	18.42
PS	13.57	92.93	66.67	0.57	0.57	40.22	52.17	7.61
SP	21.75	72.28	59.41	0.61	0.39	68.49	20.55	10.96
META	16.78	88.30	56.40	0.57	0.56	46.99	44.58	8.43

The optimality is not on the higher side. This, we argue, is because humans had no incentive to reach an optimal solution which is the case most of the time in real life. We have argued in Sect. 5 that coordination is necessary to reach optimal solutions, which in turn is attained if the negotiating parties develop their policies according to their opponent. This opens up an option for future work where AI agents actually undergo learning in real time by playing many negotiating games against humans. The issue, of course, is the scale of the training instances, which needs to be solved.

9 Conclusion and Future Work

We have introduced a simple way of using *deep reinforcement learning* to train agents for contract negotiation. As this is a multiagent setting, we have been successful in modeling varying behaviors in agents by varying the reward signal in the player and its opponent. Empirical evidence shows coordination in agents so as to reach optimal solutions. We have also built a meta-agent that shows a dynamic behavior by learning an ensemble of models. That our agents can be deployed in real-world scenarios is evidenced by the consistency in their trained behaviors against real human players. The fact that we use reinforcement learning to build agents with multiple negotiating behaviors and provide evidence for their usability with human evaluation makes this work a unique contribution. There is much scope for improvement in terms of behavior modeling wherein we may use reinforcement learning to learn hyperparameters involved in the *proposal curves* as used in Ref. [4] and also in the reward signals in Ref. [21]. We need to look into ways to train agents to ground their communication in natural language (NL) while negotiating by making them perform a parallel NL task as done in Ref. [15].

Hyperparameter Settings

The OfferMLP() is a 2-layer MLP with 64 hidden units in each layer and ReLU activation. Both AgentLookup() and TurnLookup() are embeddings with size 32. The GRU() has a hidden state size of 256. The baseline b_i and b_s is computed as the running average of the rewards received for each agent i. The entropy regularization weight parameter λ is changed in every epoch. For 5 epochs, the values of λ are [0.1, 0.05, 0.01, 0.005, 0.001]. The discount factor γ is set to 0.99.

For training, we used SGD optimizer with nesterov [20] and momentum of 0.1. The learning rate was fixed to 0.01.

References

1. R. Axelrod, W.D. Hamilton, The evolution of cooperation. Science **211**(4489), 1390–1396 (1981)
2. L.M. Camarinha-Matos, A.I. Oliveira, Contract negotiation wizard for vo creation, in *Digital Enterprise Technology* (Springer, 2007), pp. 333–342
3. K. Cao, A. Lazaridou, M. Lanctot, J. Leibo, K. Tuyls, S. Clark, Emergent communication through negotiation (2018). arXiv:1804.03980
4. M. Cao, X. Luo, X.R. Luo, X. Dai, Automated negotiation for e-commerce decision making: a goal deliberated agent architecture for multi-strategy selection. Decision Support Syst. **73**, 1–14 (2015)
5. K. Cho, B. van Merrienboer, D. Bahdanau, Y. Bengio, On the properties of neural machine translation: encoder–decoder approaches, in *Proceedings of SSST-8, Eighth Workshop on Syntax, Semantics and Structure in Statistical Translation* (2014), pp. 103–111
6. P.E. Dunne, Extremal behaviour in multiagent contract negotiation. J. Artif. Intell. Res.(JAIR) **23**, 41–78 (2005)
7. P.E. Dunne, M. Wooldridge, M. Laurence, The complexity of contract negotiation. Artif. Intell. **164**(1–2), 23–46 (2005)
8. M.S. English, P.A. Heeman, Learning mixed initiative dialog strategies by using reinforcement learning on both conversants, in *Proceedings of the conference on Human Language Technology and Empirical Methods in Natural Language Processing* (Association for Computational Linguistics, 2005), pp. 1011–1018
9. P. Faratin, M. Klein, Automated contract negotiation and execution as a system of constraints, in *Proceedings of the Workshop on Distributed Constraint Reasoning, at IJCAI*, vol. 1 (2001)
10. J. Foerster, I.A. Assael, N. de Freitas, S. Whiteson, Learning to communicate with deep multi-agent reinforcement learning, in Adv. Neural Inf. Process. Syst. 2137–2145 (2016)
11. K. Georgila, C. Nelson, D. Traum, Single-agent versus multi-agent techniques for concurrent reinforcement learning of negotiation dialogue policies, in *Proceedings of the 52nd Annual Meeting of the Association for Computational Linguistics (Volume 1: Long Papers)*, vol. 1 (2014), pp. 500–510
12. C. Guthrie, Negotiations: statistical aspects. Amer. Psychol. **39**, 341–50 (2001)
13. P.A. Heeman, Representing the reinforcement learning state in a negotiation dialogue, in *IEEE Workshop on Automatic Speech Recognition & Understanding, 2009. ASRU 2009* (IEEE, 2009), pp. 450–455
14. C. Hou, Modelling agents behaviour in automated negotiation, in *Technical Report KMI-TR-144, Knowledge Media Institute* (2004)
15. A. Lazaridou, A. Peysakhovich, M. Baroni, Multi-agent cooperation and the emergence of (natural) language. arXiv:1612.07182 (2016)
16. A. Lerer, A. Peysakhovich, Maintaining cooperation in complex social dilemmas using deep reinforcement learning. arXiv:1707.01068 (2017)
17. M. Lewis, D. Yarats, Y.N. Dauphin, D. Parikh, D. Batra, Deal or no deal? end-to-end learning for negotiation dialogues. arXiv:1706.05125 (2017)
18. V. Mnih, A. Puigdomenech Badia, M. Mirza, A. Graves, T. Lillicrap, T. Harley, D. Silver, K. Kavukcuoglu, Asynchronous methods for deep reinforcement learning, in *International Conference on Machine Learning* (2016), pp. 1928–1937
19. V. Nair, G.E. Hinton, Rectified linear units improve restricted Boltzmann machines, in *Proceedings of the 27th International Conference on Machine Learning (ICML-10)* (2010), pp. 807–814
20. Y. Nesterov, A method of solving a convex programming problem with convergence rate o (1/k2). Soviet Math. Doklady **27**, 372–376 (1983)
21. A. Peysakhovich, A. Lerer, Prosocial learning agents solve generalized stag hunts better than selfish ones. arXiv:1709.02865 (2017)
22. H. Raiffa, *The Art and Science of Negotiation* (Harvard University Press, 1982)

23. C. Sierra, P. Faratin, N.R. Jennings, A service-oriented negotiation model between autonomous agents, in *Collaboration Between Human and Artificial Societies* (Springer, 1999), pp. 201–219
24. R.J. Williams, Simple statistical gradient-following algorithms for connectionist reinforcement learning, in *Reinforcement Learning* (Springer, 1992), pp. 5–32
25. F. Yang, P.A. Heeman, Using computer simulation to compare two models of mixed-initiative, in *Eighth International Conference on Spoken Language Processing* (2004)
26. Y. Nanpeng, L. Tesfatsion, C.-C. Liu, Financial bilateral contract negotiation in wholesale electricity markets using nash bargaining theory. IEEE Trans. Power Syst. **27**(1), 251–267 (2012)
27. M. Zuzek, M. Talik, T. Świerczyński, C. Wiśniewski, B. Kryza, Ł. Dutka, J. Kitowski, Formal model for contract negotiation in knowledge-based virtual organizations, in *International Conference on Computational Science* (Springer, 2008), pp. 409–418

Effective Acceptance Strategy Using Cluster-Based Opponent Modeling in Multilateral Negotiation

Zahra Khosravimehr and Faria Nassiri-Mofakham

Abstract Determining required conditions for accepting a bid has an important role in efficient negotiations. Conceding (aggressive) acceptance strategy of the agent against a conceding (an aggressive) opponent can lead to early and high utility agreement (losing the utility) especially in domains with discount factor. The literature is more focused on bilateral than multilateral negotiations, where they assume cost or utility function of the opponents or the historical data among the negotiation sessions are known to the agents. In this study, an effective acceptance strategy is proposed which only assumes the opponents' rationality. The strategy extends the ParsAgent last moment concession strategy by developing an opponent model using *k*-means. The performance of the strategy is investigated against different combination of opponents consisting state-of-the-art strategies from Automated Negotiating Agents Competition (ANAC 2015 and 2016) in trilateral tournaments in different domains. The evaluation results indicate the superiority of proposed agent against the opponents in all domains, in average.

1 Introduction

Human communities gain consensus and resolve conflicts through complex negotiations in personal or business situations. Artificial software agents can reduce these efforts by negotiating on behalf of humans. Agents negotiate with opponents by exchanging proposals under a predetermined protocol. The set of all possible out-

Z. Khosravimehr · F. Nassiri-Mofakham (✉)
Facuty of Computer Engineering, University of Isfahan, Isfahan, Iran
e-mail: fnasiri@eng.ui.ac.ir; fnm@cwi.nl
URL: https://eng.ui.ac.ir/ fnasiri; https://www.cwi.nl/people/faria-nassiri-mofakham

Z. Khosravimehr
e-mail: khosravimehr@gmail.com

F. Nassiri-Mofakham
Intelligent and Autonomous Systems, Centrum Wiskunde & Informatica, Amsterdam, The Netherlands

T. Ito et al. (eds.), *Advances in Automated Negotiations*, Studies in Computational Intelligence 905, https://doi.org/10.1007/978-981-15-5869-6_6

comes of the negotiations is called negotiation domain. Some domains have discount factor which affects the utility with the passage of time. Each agent has some preferences on each negotiation domain that is referred to preferences profile [1, 2].

In every negotiation with a deadline, one of the negotiating parties in bilateral negotiation [6] or multiple negotiating parties in multilateral negotiation [9] must accept an offer to avoid a break off. Therefore, the use of an effective mechanism for identifying the necessary conditions for the acceptance of the proposals could lead to effective outcome [2]. Despite numerous research studies in opponent modeling and bidding strategy [5], not much work has been done in acceptance strategy, while multilateral negotiations have been less considered. Moreover, to simplify the negotiation situations the strategies use some assumptions such as availability of the information about utility function or preferences of the opponent [9], history of the negotiations conducted by the opponent in the past [11], or the same utility function for the opponent as for the agent [13]. Evaluating the first ANAC tournament showed that the agents in bilateral negotiations use simple acceptance condition regardless of the information about the opponent [3]. However, investigating the state-of-the-art agents participating in multilateral ANAC competitions (2015 and 2016) shows that they mostly use a combination of methods (such as constant or time-based utility function) in their acceptance strategy and/or using opponent information (including the number of accepted offers, bids received from the opponent, and so on). The studies also show that these agents use some assumptions like the same utility function for the opponents as for the agent. Table 1 compares the acceptance strategies of the winners of ANAC 2016 and the agents participated in ANAC 2015 with respect to AC_{next} (i.e., in accepting an offer received from the opponent, the agent considers the next offer ready to send), using constant utility function, time-dependency, using combination of methods, and using opponent information. Several state-of-the-art agents employ modeling characteristics of an opponent such as acceptance strategy, deadline, preferences profile, and bidding strategy.

Two general approaches have been used for the estimation of opponent acceptance strategy: estimating reservation value of the opponent and estimating the probability of his/her acceptance [5]. In negotiations about a measurable issue, when the conflict of the agents' preferences is a common knowledge (e.g., seller prefers higher prices, while buyers prefer lower prices), knowing the reservation value of the opponent is enough to determine all acceptable proposals. An opponent model can learn reservation value of the opponent considering the rate of his/her cooperation. Current methods for estimating reservation value are based on the fact that the agents usually stop concession near their reservation value and this behavior occurs near the end of negotiation. These models assume that domain information is available and the opponent uses a specific and predetermined strategy [5, 8, 12, 15, 16]. All of these studies are about bilateral and single-issue negotiations. The present study extends ParsAgent [10] by proposing an acceptance strategy and opponent model which is not limited to such assumptions and works in multi-issue, multilateral, and incomplete information negotiation. The only thing that it assumes is rationality of the opponent (i.e., she/he does not propose bids which brings him/her a low utility). The model estimates the opponent's current reservation value by online clustering oppo-

Table 1 Acceptance strategy of ANAC 2015 and ANAC 2016 winner agents

Agent	AC_{next}	Constant utility	Time	Combination	Using opponent info.
Atlas3	–	–	×	×	×
ParsAgent	–	–	×	×	×
RandomDance	×	–	×	×	–
Kawwaii	–	–	×	×	×
AgentBuyog[a]					
PhoenixParty	×	–	×	×	×
XianFaAgent	–	–	×	×	×
PokerFace	–	×	×	×	–
CUHK	×	–	×	×	×
Ares	×	–	×	×	×
Mercury	×	–	×	×	×
AgentH	–	×	×	×	–
SENGOKU	–	–	×	×	×
AgentW	–	–	×	×	×
AgentX	–	–	×	×	×
DrageKnight	–	×	×	×	–
XianFa	–	–	×	×	–
Neo	×	×	–	×	×
HP	–	×	×	×	×
Group2	–	–	×	×	–
TUDMixedStrategy	×	–	–	×	×
Mean	–	×	×	×	–
JonnyBlack	–	×	–	×	×
Caduceus	×	×	×	×	×
ParsCat	–	–	×	–	–
YXAgent	–	–	×	×	×

[a]Due to runtime error, this agent could not be analyzed

nent bids exchanged during a negotiation session. This study considers the fact that the opponent reservation value can be changed based on the remaining time, so that finally reaches the lowest reservation value of the domain. The proposed acceptance strategy uses the estimated reservation value for calculating the minimum acceptable utility or reservation value of the agent.

The structure of the paper is as follows: Sect. 2 presents the proposed agent, Sect. 3 describes the negotiation setting, Sect. 4 illustrates the experiment results, and Sec. 5 concludes.

2 Proposed Negotiating Agent Strategy

The proposed agent estimates opponent acceptance strategy by estimating the opponent reservation value in multi-issue multilateral incomplete information negotiation. The agent concedes against an opponent that cooperates, and then decreases his/her current reservation value to his/her minimum reservation value and quickly agrees with low-utility offers, so the agreement is reached earlier, and the utilities of the parties especially in domains with discount factor will increase. In contrast, against opponents who behave hardheaded, cooperative behavior can lead to exploitation so that the opponents get great utility but bring a little profit to the agent. Using hardheaded behavior against such opponents does not lead to exploitation, so that in the case of disagreement, it receives a profit equals to the opponent, at least. This study proposes ParsAgent_E, which extends ParsAgent [10] in two steps: (1) an opponent model using learning based on clustering is proposed, (2) its acceptance strategy is extended based on this opponent model.

2.1 *Opponent Model*

One of the challenges in negotiation with incomplete information is the lack of output variable (i.e., the variable which indicates the utility values of the received bids in the opponent's view). Therefore, specific opponent model which can model the reservation value of the opponent without any need for the value of the output variable is required. It sounds that models with unsupervised learning techniques are the only way to model opponent behavior without any restricting assumptions.

We define concession of each opponent at any given time to be the utility that the opponent refuses. The reservation value at any time equals to the minimum acceptable utility for the opponent at that time, that is, the amount of the utility that the opponent does not refuse. Therefore, the relationship between the opponent's reservation value $R(op)$ and the opponent's concession $C(op)$ at any time is $R(op) = 1 - C(op)$. This equation is applied to calculate temporal reservation value of each opponent at each moment. Clustering technique is used to determine the opponent's concession. Due to time, memory, and processor limitations in Genius framework, we should use a simple and fast clustering method that does not consume too much memory. The proposed model does not make any assumption about the opponent behavior except its rationality in the sense that she/he does not insist on proposing offers having low utility. The idea behind the approach notices the fact that the high distribution of the opponent's bids shows that he/she proposes bids with different utilities and so his/her reservation value is low; because lower reservation value obviously leads to submitting a wider range of proposals. The compression of proposals means the opponent has participated more aggressively in negotiation and in fact she/he has a higher reservation value. In other words, the more offer dispersion, the more utility dispersion, the less concentration around single offers, the lower reservation value, while the less dispersion, the more boulware behavior, the less conssession, the higher reservation value. Belonging to a specific cluster is not very important and only dispersion or compaction of clusters are important for estimating reservation value. This also holds for single-issue domains. For this reason, the famous and simple k-means clustering method is used. The number of clusters, k, the algorithm applies is 20, which was determined through a few limited experiments (using $k = 3, 5, 10, 20, 30$). Since clustering all the received bids, which in some cases are several hundred thousand offers, needs more time and memory than the limits set for the tournament, so clustering is done in two steps. First, it clusters every 100 offers received from the opponent. Next, clustering is performed on cluster centers (centroids) obtained from the first step. Distance of the bid from cluster centers is calculated using Eq. (1) and the offer is placed in a cluster with the smallest distance between the bid and the center of the cluster.

$$d(p, \mathrm{central}_j) = \sum_{i=0}^{n} \begin{cases} 1, & \mathrm{val}(i)_{\mathrm{center}_j} = \mathrm{val}(i)_p \\ 0, & \mathrm{val}(i)_{\mathrm{center}_j} \neq \mathrm{val}(i)_p \end{cases} \tag{1}$$

In Eq. (1), n is the number of issues or domain issues, p is the received bid, j is the cluster number, $\mathrm{central}_j$ proposed center of the cluster and function $\mathrm{val}(i)$ returns the value of the issue i for a bid. Here the value of the issue for central bid is compared with the proposed offer. If equal, it returns 1, otherwise it returns zero. And $d(p, \mathrm{central}_j)$ calculates the distance of p from the center of cluster j. It is worth noting that all domains used in ANAC 2015 have discrete issues. This equation could be generalized for estimating the distance for continuous values. Finally, two parameters *non-uniform distribution* and *the maximum distance of cluster centers* are extracted from the clusters. These parameters play an important role in determining the reservation value of the opponent. The greater the maximum distance of the cluster centers, the greater the degree of cooperation and in fact the less the amount of reservation value. On the other hand, the greater non-uniform distribution parameter means the opponent acts hardheaded and has higher reservation value as described below.

First, the distribution of offers in clusters is examined, how far the bids are distributed across clusters or concentrated in a cluster. The non-uniform distribution of the cluster created for the bids received from the opponent op by $dist(op)_{abnormal}$ is defined as

$$\mathrm{dist(op)}_{\mathrm{abnormal}} = \sum_{i=1}^{k} \left| \mathrm{dist}_{\mathrm{normal}} - \frac{\mathrm{member}_i}{N} \right| \tag{2}$$

where, k is the number of clusters, $member_i$ is the number of members in cluster i, N is the total number of clustered bids for opponent op and k shows the number of clusters. So $member_i/N$ shows the distribution percentage in cluster i. The greater this distribution parameter indicates that the opponent has insisted on giving the same bids and she/he has acted stubbornly.

Maximum distance of cluster centers max _dist(op) is the second parameter that is obtained from clustering results. Each time the distance of all cluster centers is calculated for each opponent, the maximum amount of this distance is considered as the value of this parameter defined by

$$\max\_\text{dist}(\text{op}) = \max\left(d(\text{central}_i, \text{central}_j)\right), \forall clusters i, j \quad (3)$$

Finally, to map the values of these parameters to a number between zero and 1, the value of the max _dist parameter is divided by total number of domain issues. So this parameter does not depend on the domain size.

Also, $\text{dist(op)}_{\text{abnormal}}$ is divided by the maximum non-uniform distribution value which is equal to

$$\max\_\text{dist}_{\text{abnormal}} = 1 + \frac{1}{k}(k-2) \quad (4)$$

where, k determines the number of clusters and $\frac{1}{k}$ shows uniform distrbution of the bids in the cluster. Since the largest non-uniform distribution is related to the time when all the proposals fall into a cluster and the other clusters do not have any members, therefore the value of $\text{dist}_{\text{abnormal}}$ for the cluster containing all the members is equal to $(|\frac{1}{k} - 1|)$ and for the other clusters is equal to $(\text{dist}_{\text{normal}} - 0)$.

For simplicity, we consider a linear relation between these parameters, *non-uniform distribution* and the *maximum distance of cluster centers*, and the reservation value. So, the amount of concession of each opponent will be in the form of

$$C(\text{op}) = w_1 \frac{\max\_\text{dist(op)}}{n} - w_2 \frac{\text{dist(op)}_{\text{abnormal}}}{\max\_\text{dist}_{\text{abormal}}} \quad (5)$$

where, w_1 and w_2 are the weights of the parameters and L is the number of negotiated domain issues. This study considers constant values 0.6 for w_1 and 0.4 for w_2 (other values could be considered in the future work). The algorithm used in these two phases is proposed below. Briefly, per each opponent, centroids of each 100 last bids are determined and added to the existing list of centroids. So, there would be, more than 20 centroids at the end of negotiation. Among all, Algorithm 2 chooses centroids and updates them.

Algorithms 1 and 2 are executed for each opponent and run for every 100 received offers. In Algorithm 1, 100 new bids received from the opponents are clustered and the cluster centers are identified. Then, in Algorithm 2, all the cluster centers for each opponent are clustered again and the *non-uniform distribution* and the *maximum distance of cluster centers* parameters are calculated, yielding the opponent's reservation value.

2.2 Acceptance Strategy

Proposed strategy uses a time-dependent target utility function

Algorithm 1: Opponent Concession Rate: Level 1

Input : 100 newest opponent's offers, number of clusters k
Output: clustering 100 offers and determining centroids
Begin :

```
Define 1, (100/k − 1), 2 × (100/k − 1), ···, 19 × (100/k − 1) opponent bids as initial
centroids for k clusters;
while any centroid is updated do
    Cluster opponent bids;
    Calculate the distance according to Eq. (1) between each pairs of offers in a cluster and
    select the offer with the minimum distance as new centroid;
    Add new centroids of clusters for each opponent to its ListOfCentroids.
```

Algorithm 2: Opponent Concession Rate: Level 2

Input : ListOfCentroids of opponent's bids, number of clusters k
Output: opponent's reservation value
Begin :

```
Define 1, (100/k − 1), 2 × (100/k − 1), ···, 19 × (100/k − 1) from ListOfCentroids as
initial centroids for k clusters;
while any centroid is updated do
    Cluster bids in ListOfCentroids;
    Calculate distance according to Eq. (1) between each pairs of offers in a cluster and
    select the offer with the minimum distance as new centroid;
    Calculate max_dist(op) and dist(op)_abnormal parameters and then C(op) according to
    Eqs. (2), (3) and (5);
    return R(op) = 1 − C(op) as opponent's reservation value.
```

$$u_T(t) = P_{\min} + (P_{\max} - P_{\min})\big(1 - F(t)\big) \tag{6}$$

where t is time and

$$F(t) = k + (1 - k)t^{1/\beta}. \tag{7}$$

In Eq. (6), $P_{\min}$ and $P_{\max} \in [0, 1]$ determine the maximum and minimum values of u_T function in the interval 0 and 1. The u_T function here is used as acceptable utility for the agent starting from $P_{\max}$ and ending with $P_{\min}$. The parameter $k \in [0, 1]$ represents the utility at the zero time or the beginning of the negotiation. In many cases, for simplicity, the value of $P_{\min}$ is zero, $P_{\max}$ is 1 and k is equal to 0. In this way, a simple relationship is obtained that ParsAgent has also used in its acceptance strategy. In Eq. (7), β is the parameter that controls the speed of the concession of the agent in the negotiation. When β is equal to, less, or greater than 1, the behavior of the agent will be linear, hardheaded, or cooperative. ParsAgent acceptance conditions use this strategy with $\alpha = 0.7$ and $\beta = 0.2$, but $\beta = 0.15$ in domains without discount factor. The acceptance strategy of the agent is according to

$$\mathrm{AC}_{\mathrm{combi}}(T, \alpha) \Leftrightarrow \big(U_A(x^t_{B\to A}) \geq u_T(t)\big) \wedge \big(U_A(x^t_{B\to A}) \geq \alpha\big)$$

where, $x^t_{B \to A}$ is the offer received from the opponent at (current) time t, T (deadline) is the time the negotiation will take from the beginning, and the constant α is the minimum acceptable value. Using a constant value in ParsAgent at the end of the negotiation to compare the utility of the opponent's offer and deciding whether to accept or reject it, could lead to inefficiencies in some circumstances. Also, the same condition may hold for β. How aggressive or cooperative the agent behave would adapt with against different opponents and in different domains. In the proposed strategy, after determining the reservation value of the opponents at any moment, the average of the opponents' reservation values is considered as the policy for determining the temporal reservation value of the agent. In addition to the reservation value, the slope of β in Eq. (7) would be faster and slower against the cooperative and hardheaded opponents, respectively. In this way, the slope of the equation and the reservation value have a converse relation. By decreasing the reservation value of the agent, the velocity and gradient of β graph will be increased to reach an agreement faster.

With considering the maximum utility $P_{\max} = 1$ in Eq. (6) for the proposed agent, the acceptance strategy of the proposed agent is a hybrid strategy in the form of

$$AC_{combi}(T, P_{\min}, \beta) \Leftrightarrow \left(U_A(x^t_{B \to A}) \geq u_T(t, P_{\min}, \beta)\right)$$

where $P_{\min}$, which specifies the minimum acceptable utility or the reservation value of the agent, and β are determined during the negotiation from

$$u_T(t, P_{\min}, \beta) = P_{\min} + (1 - P_{\min})(1 - t^{1/\beta}) \tag{8}$$

and the following steps: 1) at the beginning of the negotiation, the acceptance strategy of the agent is the same as the hardheaded acceptance strategy of ParsAgent, but $P_{\min}$ and β are respectively 0.8 and 0.2 to gain a great utility as much as possible in the case of an agreement, 2) the algorithm for calculating the reservation value of each opponent is performed periodically. If the generated reservation value for the agent is less than the previous $P_{\min}$ value calculated in the previous round or is less than 0.8 (if the first round), that is the opponent behaves cooperatively and so this value is replaced with the previous value as well as the calculated concession amount for β. Otherwise, the default hardheaded behavior continues.

In the proposed acceptance strategy, by taking into account the fact that many agents behave cooperative in the final moment of the negotiation to prevent a break off, at one second remained to the end of the negotiation, the minimum acceptable utility minus 0.1 is acceptable and in 0.2 seconds remaining to the end of the negotiation, the minimum utility will be acceptable for the proposed agent.

3 Multilateral Negotiation Setting

To evaluate the acceptance strategy of the proposed agent and compare with other state-of-the-art acceptance strategies of ANAC 2015 and 2016 competitions, it is necessary to set up a negotiation framework where they have no bidding component to have no impact on the negotiation outcome. However, for evaluating all the agents in the same conditions, the opponents require to behave in non-adaptive manner and just bid independent of the strategy of the participating agents. Among all, time-dependent strategies are the most famous non-adaptive ones used for opponent agents in such experiments where time is the only effective factor for these agents in choosing a bid to send [4, 14]. Therefore, this setting employs four groups of opponents with different time-dependent bidding behaviors, as follows:

1. *Conceding agent* applies a time-dependent concession strategy to generate bids during the negotiation session, so that the target utility is calculated through $u_T(t) = P_{\max}(1 - t^{1/\beta})$, where, 0.2, 1, and 2 are applied for concession rate β with a fixed maximum bid utility $P_{\max}$ equal to 1. Over the time, the agent satisfies even with lower utility bids. Acceptance strategy of the agent is $\mathrm{AC}_{\mathrm{next}}$; that is, a received bid is accepted when its utility is greater than or equal to the utility of the offer ready to send.
2. *Random agent* generates a bid with a utility above a fixed threshold m in a random manner, where $m = 0, 0.5, 0.7$. Acceptance strategy of the agent uses the same threshold to accept or reject the opponent's offer. When the utility of the received bid is higher than this threshold, the bid will be accepted.
3. *Conceding agent with an offset* is the same as conceding agent but does not start the negotiation with the best bid. The agent applies a linear concession rate $(\beta = 1)$ and maximum bid utility $P_{\max} = 0.7, 0.8, 0.9$.
4. *Non-conceding agent* starts with a minimum target utility which increases toward its maximum over the time. Acceptance strategy of the agent is AC_{next}. The target utility is calculated through $u_T(t) = P_{\min} + (1 - P_{\min})t$, where the minimum acceptable utility $P_{\min} = 0, 0.25, 0.75$.

In trilateral negotiation, the agent negotiates with two non-adaptive opponents, that is 78 combinations of these 12 benchmark opponents are used in the experiment setting. The opponents ignore bids received from the agent and offer a new bid. If the agent accepts the bid, the opponent will consider the offer. Therefore, the bidding strategy of the agent will have no effect on the outcome of the negotiation. But, the agent still can take advantage of the opponent model in the acceptance strategy. The experiments run in 10 different domains used in ANAC 2015 [7], where they are illustrated in Table 2.

The study performs two experiments for determining the *accuracy* and *performance* of the proposed opponent model and acceptance strategy against opponent agents. Since the proposed model estimates the reservation value of the opponent, the error described as Eq. (9) which determines the distance between the estimated value x and the real value x' is used as the accuracy criterion [2].

Table 2 Experimental domains

No.	Size (issues, outcomes)	Discount factor	Reservation value	Competitiveness
1	small (1, 5)	0	0.5	Very cooperative
2	small (1, 5)	0	0.5	A little cooperative
3	small (2, 25)	0.2	0	Very competitive
4	small (2, 25)	0.1	0.5	Extremely collaborative
5	medium (4, 320)	0.5	0	Competitive
6	medium (4, 320)	0.5	0	Cooperative
7	large (8, 3^8)	0	0	Competitive
8	large (8, 3^8)	0	0	Cooperative
9	very large (16, 3^{16})	0.4	0.7	Very cooperative
10	very large (16, 3^{16})	0.4	0.7	Very competitive

$$d(x, x') = |x - x'| \tag{9}$$

The performance of the proposed strategy compared with real agents is evaluated via *average utility* as the main criterion used in ANAC competitions to identify the rank of the agents in the individual utility category [3] according to

$$\text{AvgUtility} = \frac{\sum_{i=1}^{N} \text{utility}_A^i}{N} \tag{10}$$

where, N is the total number of negotiation sessions, AvgUtility_A is the average utility obtained for agent A in N sessions, and utility_A^i is the utility of agent A in the ith negotiation session.

4 Experimental Results

4.1 Experiment I: Accuracy

This experiment determines the accuracy of the proposed opponent model by calculating its error in estimation of the reservation value for every 100 bids received from each of 12 benchmark opponents in 10 different negotiation domains according to Eq. (9). Averaged per each of four opponent types, Fig. 1 shows that the estimated error against all the opponents in small domains is greater than larger domains. The model is more error-prone in dealing with non-conceding opponents who unusually start with low-utility bids but continue with higher utility bids. Also, the accuracy of the model does not have much relation with other domain features such as discount factor and reservation value.

4.2 Experiment II: Performance

The performance of the proposed strategy is compared with 23 participating agents in ANAC 2015 including ParsAgent (Buyog agent was not used due to an error during execution) and top 3 ANAC 2016 agents state-of-the-art agents using average utility (Eg. 10) in 78 tournaments each in 3 minutes against combinations of 12 benchmark opponent pairs each with 6 different profiles in 10 negotiation domains, repeated 3 times to be valid.

As Fig. 2 illustrates, the proposed agent has obtained relatively good performance in all domains. Table 3 summarizes the utility values earned by the proposed agent versus the best agent in each domain. As the data in Table 3 shows, proposed agent performs better in large than small domains, because the clustering function can be applied more effectively in these domains since it has small error in estimating the reservation value in large domains as shown in Fig. 1. Moreover, taking into account the constant number of 20 clusters in the proposed strategy is not appropriate for small domains, and it seems that the number of clusters would depend on the domain size. Also, the proposed agent performs better in cooperative than competitive domains. The preference profiles of the agents involved in the negotiation in the cooperative domains are not in conflict and so the possibility of offering high-value bids for all agents increases. Therefore, due to considering the average of the reservation values of the opponents as the reservation value of the agent, this policy appears to be more effective in cooperative than competing domains, as it could lead to disagreement in many negotiations in competitive domains. Hence, considering more conceding policy on these domains may lead to more utility values for the agent. On the other hand, in the domains which have a discount factor but the reservation value is zero, the proposed strategy has a weakness and it seems that applying the discount factor in the proposed strategy, especially when the reservation value of the domain does not exist, should be considered more effectively.

Table 3 summarizes the average utility values earned by the proposed agent versus the average utility of the best agent in each domain. By summing up the results, the average utility obtained by the proposed agent in all domains is shown in Table 4. Although the proposed agent (ParsAgent_E) has not been the best in any domain (see Fig. 2 and Table 1), but on average in all domains, it has gained the best result compared with the state-of-the-art agents. Also, in all domains, the efficiency of the proposed agent is much better than ParsAgent which uses a fixed value in his/her acceptance strategy and strongly depends on the type of the opponents and the domain of negotiation, while the opponents lack the opponent model and offer their bids according to time and do not repeat them. Therefore, the hardheading behavior of ParsAgent did not lead to their cooperation, and thus ParsAgent has made less utility.

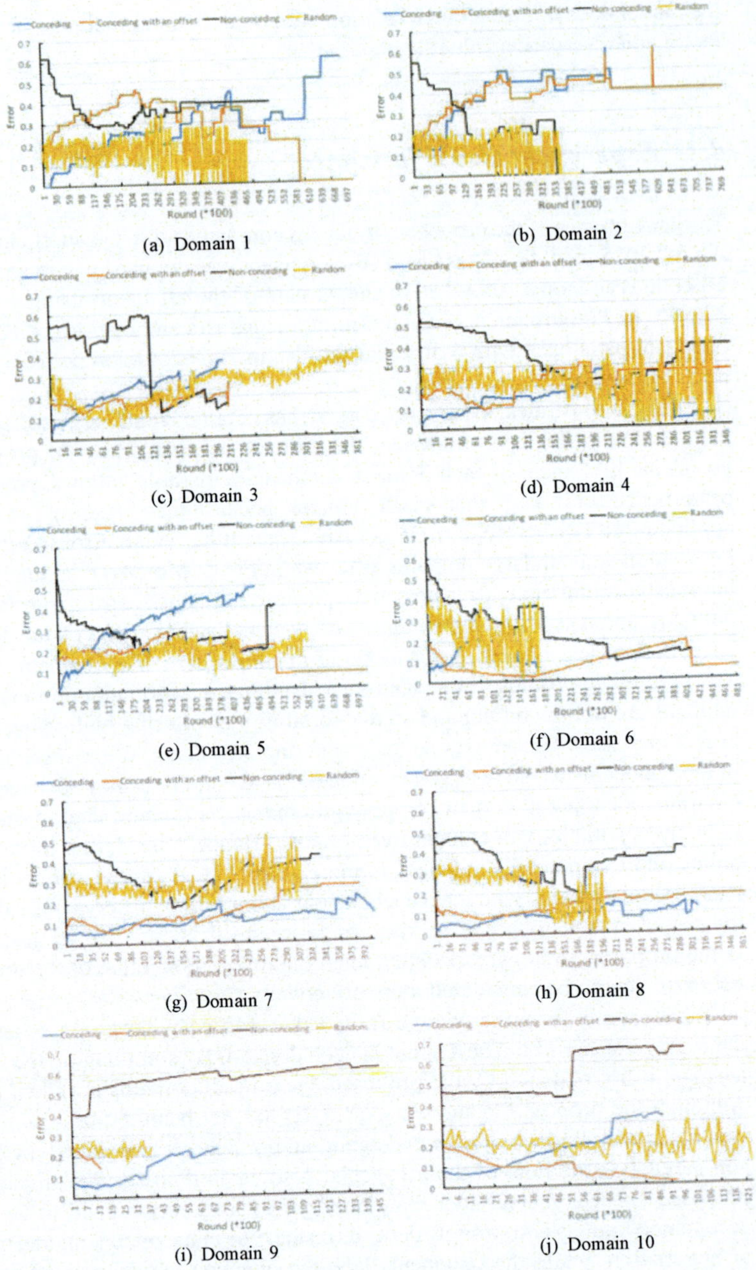

Fig. 1 The error of predicting reservation values of conceding, non-conceding, and random opponent agents in 10 different negotiation domains

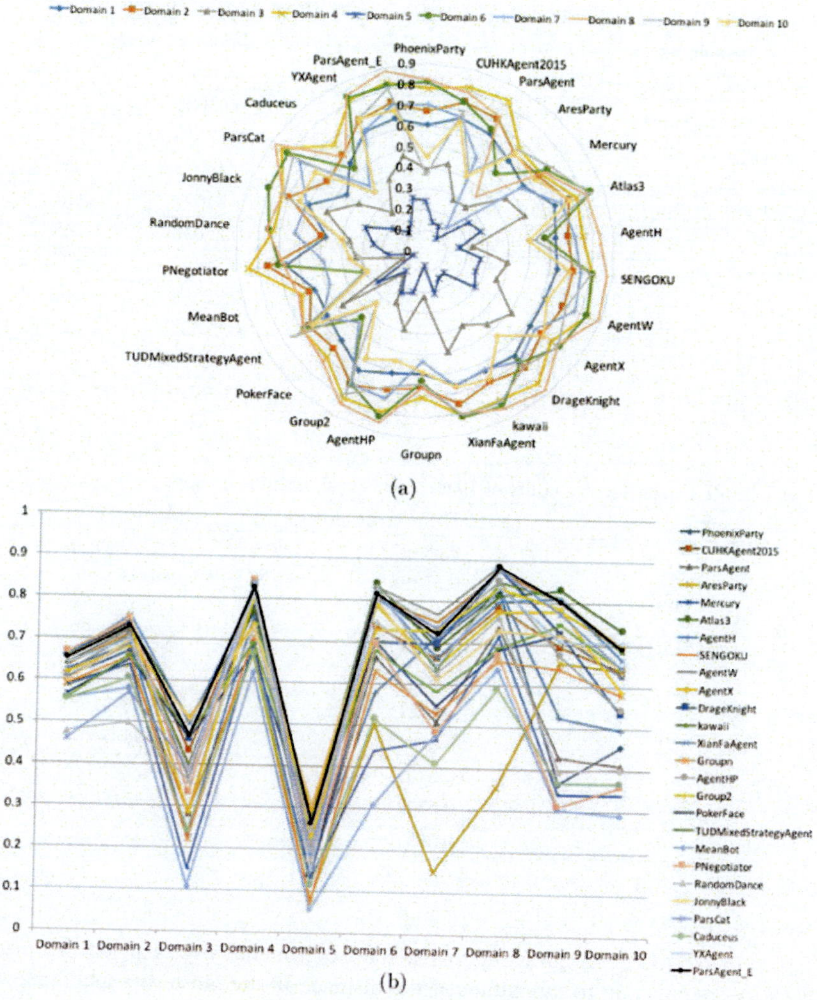

Fig. 2 The utility obtained, **(a)** in each domain by 26 agents, **(b)** per each agent in 10 negotiation domains

4.3 Key Findings

Implementation of the opponent modeling using clustering technique without the need for a specific assumption regarding the opponents or the domain, and using the technique of concession in the last moments showed a suitable method for acceptance strategy and has improved the overall efficiency of the agent. It is also worth noting that these results were performed on 10 domains with discrete issues that were used

Table 3 Performance of the proposed agent in 78 trilateral negotiations

Domain No.	Best		Proposed Agent	
	Agent	Utility	Utility	Rank
1	PNegotiator	0.66624	0.65086	7
2	XianFaAgent	0.75492	0.72856	6
3	AgentX	0.51052	0.46775	6
4	PNegotiator	0.84399	0.82627	5
5	JonnyBlack	0.30780	0.26036	6
6	Atlas3	0.83826	0.81407	7
7	AgentW	0.76882	0.72058	6
8	AgentW	0.88757	0.88301	2
9	Atlas3	0.82943	0.80060	4
10	Atlas3	0.73395	0.69016	5

Table 4 The average utility of 10 top agents in all negotiation domains

Rank	The average utility	Agent
1	0.68422	ParsAgent_E
2	0.68398	Atlas3
3	0.67457	SENGOKU
4	0.66878	XianFaAgent
5	0.66205	AgentW
6	0.65215	AgentX
7	0.64914	Kawaii
8	0.64015	YXAgent
9	0.63711	Group2
10	0.63579	DrageKnight

in ANAC 2015 competitions. If the experiments were carried out in continuous domains, due to the nature of the distance in the clustering technique, the results could be much better. This is something that will be addressed in future research. Briefly, the following key properties and findings obtained:

Properties: 1) The proposed method is more accurate in larger domains, 2) The proposed method works better in larger domains, 3) The proposed method has generally been better in cooperative domains, 4) The concession behavior of the proposed agent at the last moments decreases the number of disagreements and increases the utility of the agent, 5) Since the proposed method has been relatively weak in the domains with discount factor with zero reservation value, the number and time of concessions seem to be a function of the discount factor and the reservation value of the domain, 6) Since the acceptance strategy of extended ParsAgent uses fixed value, it yields a low utility value against various opponents with different strategies, as the opponents used in the experiment do not repeat their bids at different intervals and lack the opponent model.

Findings: 1) Using techniques that require fewer assumptions is usable and efficient in a wider range of opponents and domains, 2) Models that use certain assumptions are generally very good if the conditions are being held, but weak if they fail.

5 Conclusions and Future Works

In this paper, an acceptance strategy was proposed using an effective opponent model that considers no hypothesis except the rationality of opponent. The proposed acceptance strategy develops an opponent model using k-means clustering method and concedes at the last moments of the negotiation. The effectiveness of this strategy was tested via experiments performed against state-of-the-art opponent agents in 78 tournaments in 10 different negotiation domains. The proposed agent reached the best results in all 10 domains, in average, although not at top in any domain. Therefore, the proposed strategy has the capability to combine with an efficient bidding strategy in the BOA framework [2] and provide a powerful agent for negotiating against various opponents in different domains. The study considers 20 clusters. Using dynamic number of clusters and finding the relation to the type of domains can be considered to improve the proposed approach. Another interesting study could devise and compare different models by using several clustering techniques. To determine the minimum acceptable utility for the agent or reservation value of the agent, the average of the estimation of reservation values of the opponents was used. This policy can also be changed with respect to the type of the opponents or the domain. Another issue is to examine the efficiency and accuracy of the model in continuous domains. Moreover, investigating the applicability of the proposed approach in scenarios with highly non-linear utility function would be valuable.

References

1. R. Aydoğan, C.M. Jonker, K. Fujita, T. Baarslag, T. Ito, R. Hadfi, K. Hayakawa, A baseline for nonlinear bilateral negotiations: the full results of the agents competing in anac 2014, in *Current and Future Developments in Artificial Intelligence*. Bentham Science (2017), pp. 93–121
2. T. Baarslag. *What to bid and when to stop*. PhD thesis, Delft University of Technology (2014)
3. T. Baarslag, K. Fujita, E.H. Gerding, K. Hindriks, T. Ito, N.R Jennings, C. Jonker, S. Kraus, R. Lin, V. Robu, et al., Evaluating practical negotiating agents: results and analysis of the 2011 international competition. Artif. Intell. **198**, 73–103 (2013)
4. T. Baarslag, M. Hendrikx, K. Hindriks, C. Jonker, Predicting the performance of opponent models in automated negotiation, in *Proceedings of the 2013 IEEE/WIC/ACM International Joint Conferences on Web Intelligence (WI) and Intelligent Agent Technologies (IAT)-Volume 02*. IEEE Computer Society (2013), pp. 59–66

5. T. Baarslag, M.J.C. Hendrikx, K.V. Hindriks, C.M. Jonker, Learning about the opponent in automated bilateral negotiation: a comprehensive survey of opponent modeling techniques. Autonom. Agents Multi-Agent Syst. **30**(5), 849–898 (2016)
6. T. Baarslag, K.V. Hindriks, Accepting optimally in automated negotiation with incomplete information, in *Proceedings of the 2013 International Conference on Autonomous Agents and Multi-agent Systems*. International Foundation for Autonomous Agents and Multiagent Systems (2013), pp. 715–722
7. K. Fujita, R. Aydoğan, T. Baarslag, K. Hindriks, T. Ito, C. Jonker, The sixth automated negotiating agents competition (anac 2015), in *Modern Approaches to Agent-based Complex Automated Negotiation* (Springer, Berlin, 2017), pp. 139–151
8. S. Ji, C. Zhang, K.-M. Sim, H. Leung, A one-shot bargaining strategy for dealing with multifarious opponents. Appl. Intell. **40**(4), 557–574 (2014)
9. D. de Jonge, C. Sierra, Nb3: a multilateral negotiation algorithm for large, non-linear agreement. Autonom. Agents Multi-Agent Syst. **29**, 896–942 (2015)
10. Z. Khosravimehr, F. Nassiri-Mofakham, Pars agent: hybrid time-dependent, random and frequency-based bidding and acceptance strategies in multilateral negotiations, in *Modern Approaches to Agent-Based Complex Automated Negotiation* (Springer, Berlin, 2017), pp. 175–183
11. Y. Oshrat, R. Lin, S. Kraus, Facing the challenge of human-agent negotiations via effective general opponent modeling, in *Proceedings of The 8th International Conference on Autonomous Agents and Multiagent Systems*, Vol. 1. International Foundation for Autonomous Agents and Multiagent Systems (2009), pp. 377–384
12. K.M. Sim, Y. Guo, B. Shi, Blgan: Bayesian learning and genetic algorithm for supporting negotiation with incomplete information. IEEE Trans. Syst. Man Cybern. Part B (Cybern.) **39**(1), 198–211 (2009)
13. C.R Williams, V. Robu, E.H. Gerding, N.R. Jennings, Using gaussian processes to optimise concession in complex negotiations against unknown opponents, in *IJCAI Proceedings-International Joint Conference on Artificial Intelligence*, Vol. 22 (2011), p. 432
14. F. Zafari, F. Nassiri-Mofakham, Popponent: highly accurate, individually and socially efficient opponent preference model in bilateral multi issue negotiations. Artif. Intell. **237**, 59–91 (2016)
15. D. Zeng, K. Sycara, Bayesian learning in negotiation. Int. J. Human-Comput. Stud. **48**(1), 125–141 (1998)
16. M. Zhang, Z. Tan, J. Zhao, L. Li, A Bayesian learning model in the agent-based bilateral negotiation between the coal producers and electric power generators, in *International Symposium on Intelligent Information Technology Application Workshops (IITAW'08)*. IEEE (2008), pp. 859–862

Negotiating Agents Competition, Tools and Evaluation Metrics

ANAC 2017: Repeated Multilateral Negotiation League

Reyhan Aydoğan, Katsuhide Fujita, Tim Baarslag, Catholijn M. Jonker, and Takayuki Ito

Abstract The Automated Negotiating Agents Competition (ANAC) is annually organized competition to facilitate the research on automated negotiation. This paper presents the ANAC 2017 Repeated Multilateral Negotiation League. As human negotiators do, agents are supposed to learn from their previous negotiations and improve their negotiation skills over time. Especially, when they negotiate with the same opponent on the same domain, they can adopt their negotiation strategy according to their past experiences. They can adjust their acceptance threshold or bidding strategy accordingly. In ANAC 2017, participants aimed to develop such a negotiating agent. Accordingly, this paper describes the competition settings and results with a brief description of the winner negotiation strategies.

R. Aydoğan (✉)
Department of Computer Science, Özyeğin University, Istanbul, Turkey
e-mail: reyhan.aydogan@ozyegin.edu.tr

K. Fujita
Tokyo University of Agriculture and Technology, Tokyo, Japan
e-mail: katfuji@cc.tuat.ac.jp

T. Baarslag
Centrum Wiskunde & Informatica (CWI), Amsterdam, The Netherlands
e-mail: T.Baarslag@cwi.nl

R. Aydoğan · C. M. Jonker
Interactive Intelligence Group Delft University of Technology, Delft, The Netherlands
e-mail: C.M.Jonker@tudelft.nl

T. Ito
Nagoya Institute of Technology, Nagoya, Japan
e-mail: ito.takayuki@nitech.ac.jp

T. Ito et al. (eds.), *Advances in Automated Negotiations*, Studies in Computational Intelligence 905, https://doi.org/10.1007/978-981-15-5869-6_7

1 Introduction

In recent years, much attention has been paid to automated negotiation in which intelligent agents negotiate on behalf of their users. It is a fruitful research area consisting of a variety of research challenges such as preference elicitation and representation [3], reasoning on incomplete [4] or complex preferences [25, 29], designing and developing efficient negotiation strategies [10] and protocols [21], making decision under uncertainty [20], learning opponents' interest/preferences/ strategies [3, 11] and so on. In order to bring together the researchers from negotiation community and to provide benchmarks for evaluation of the proposed negotiation strategies, an annual international competition on automated negotiation namely ANAC [22] has been organized for several years in conjunction with the most prestigious AI conferences such as the International Joint Conference on Autonomous Agents (AAMAS) and Multi-Agent Systems and the International Joint Conference on Artificial Intelligence (IJCAI). As the main organizers of this competition, our aim is to introduce new research challenges every year. Up to date, the focus was mostly on bilateral multi-issue negotiation. Recently, multilateral negotiation draws the attention; consequently, ANAC 2017 encouraged the participants to pursue developing effective negotiation strategies for repeated multilateral negotiations.

In comparison with bilateral negotiation, multilateral negotiations are more complex due to their nature—requiring to establish a joint consensus among more stakeholders with varying preferences [2]. In the competition, agents followed the Stacked Alternative Offer Protocol(SAOP) in which negotiating parties exchange their bids in a turn-taking fashion till reaching an agreement or reaching the predefined deadline [1]. The challenge in the competition is not only to negotiate with multiple opponents but also to learn from previous negotiations with the same opponents. In other words, the main goal is to design effective negotiation strategies (i.e., bidding, opponent modeling and acceptance strategies) when negotiating repeatedly with the same agents in a multilateral setting.

When an agent negotiates with its opponents first time, it may not have any idea about their strategies and preferences. Therefore, it can try to model their opponents by analyzing the current bid exchanges during the undergoing negotiation. However, when the agent negotiates with the same opponents repeatedly, it can act more strategically by exploiting its previous negotiation experiences (e.g., opponent modeling about preferences/strategies). Furthermore, since the agents will negotiate multiple times, negotiating like a Hardliner (i.e., being selfish) may cause your opponent not to concede next time; therefore, on average the agent may not gain high utility. Without doubt, it depends on to what extent the opponents take previous negotiation experiences into account in their strategies. Therefore, determining to what extent collaborative behavior to be adopted is another challenge in such a negotiation setting.

ANAC 2017 has been organized in conjunction with the IJCAI 2017 in Melbourne. 18 teams from 9 different institutions participated in the competition. Agents negotiate with the same opponents on the same negotiation scenarios five times and they can access historical data from their past negotiations. That is, they know their utility distribution over the exchanged offers in any previous negotiation session according to its own utility space and previous agreements they reached. There were two winner categories: individual utility gain and social welfare in terms of product of utilities. This paper presents competition setup and the evaluation of the ANAC 2017 agents elaborately.

The rest of the paper is organized as: Sect. 2 explains negotiation setup while Sect. 3 presents the results of the competition in detailed. Related work is provided in Sect. 4. Lastly, we conclude the paper with the potential research challenges for upcoming competitions in Sect. 5.

2 Negotiation Setup

In the competition, three agents negotiate on multiple issues to come up with an agreement. Each issue $j \in I$ can take a value v_j from a predefined set of valid values for that issue denoted by D_j (i.e., $v_j \in D_j$). Each agent can access this domain information. A bid $b = (b_1, \ldots, b_{|I|})$ is an assignment of values to all issues where $b_1 \in D_1$.

Preferences of negotiating parties are represented by means of additive utility function as shown in Eq. 1 where $V_a(v_j)$ denotes agent a's valuation of the value for the issue j in the given bid and the weights of that issue is represented by $w_{a,j}$. In other words, agents sum up their weighted valuation of each issue value to calculate the overall utility. It is worth noting that it is forbidden to access opponents' preferences during the negotiation. Each agent can only access their own preferences.

$$u_a(b^t) = \sum_{j \in I} V_a(b_j^t) \cdot w_{a,j} \tag{1}$$

The interaction among agents is governed by the Stacked Alternative Offers Protocol (SAOP) [1]. According to this protocol, each agent i is expected to provide an action a_i in its turn. Type of an action is denoted as $type(a)$ which can take $type(a) = \{A^{\text{bid}}, A^{\text{accept}}, A^{\text{end}}\}$; that is, the agent can make a counter-offer, accept the offer in the negotiation table or walk away from the negotiation, respectively.

Following SAOP, the first agent starts the negotiation with an offer that is observed by all others immediately. Whenever an offer is made the next party in line can take the following actions:

- Make a counter-offer (thus rejecting and overriding the previous offer)
- Accept the offer
- Walk away (e.g., ending the negotiation without any agreement).

This process is repeated in a turn-taking clock-wise fashion until reaching an agreement or reaching the deadline. Negotiation is also terminated if one of the agents walks away. To reach an agreement, all parties should accept the offer. If no agreement has been reached at the end of the negotiation, the negotiation fails and

Table 1 Negotiation Scenarios

Name	# of issue	# of values (1)	# of values (2)	# of values (3)	# of values(4)	# of values (5)	Scenario size
GeneJack	4	10	10	10	10		10,000
MyDomain	5	4	10	8	14	9	40,320
Music	4	9	4	7	6		1512
SuperMarket	5	8	6	5	6	4	5760
Movietime	4	6	2	3	4		144
Lunch Time	3	5	4	4			80
Taxung	4	4	5	3	6		360
SmartGrid	3	5	4	4			80

agents receive their reservation utility. If an agreement on the bid b is reached, the utility of the agent i is estimated as explained above (see Eq. 1).

In the competition, the deadline is set to three minutes and each negotiation session is repeated five times. Agents are allowed to access provided historical data from their past negotiations. The historical data involves the utility distribution of the exchanged offers in previous negotiation sessions according to agent's own utility space and previous agreements if there exists. Note that agents were not allowed to keep the entire history of exchanged bids in their past negotiation. However, they can process the bid exchanges during their ongoing negotiation and model their opponent's preferences and strategies. The opponent model can be saved and updated over time.

There were 18 submissions from nine institutions:

- Maastricht University—NL
- Nagoya Institute of Technology—Japan
- Özyegin University—Turkey
- Southwest University—United States
- Technical University Of Crete—Greece
- Tianjin University—China
- Tokyo University of Agriculture and Technology—Japan
- University of Isfahan—Iran
- University of Southampton—UK.

Participation to this competition requires submitting a negotiation scenario consisting of three conflicting preference profiles with domain description. In order to evaluate the performance of the agents, We picked eight negotiations scenarios with varying domain size from the submitted negotiation scenarios. Table 1 shows number of issues, the number of possible values for each issue and size of outcome size, respectively.

In the qualification round, finalist agents are determined. Since there were 18 submissions, running whole tournament involving 18 agents in eight domains with

5 repetitions were not feasible within the given time. Therefore, two pools were generated randomly as follows:

- **Pool-1**: Farma17, ParsCat2, taxibox, Mamenchis, MadAgent, AgentKN, PonPokoAgent, SimpleAgent, Imitator
- **Pool-2**: AgentF, GeneKing, Mosa, CaduceusDC16, ParsAgent3, TucAgent, Group3, Gin, Rubick.

Genius 7.1.4 [26], automated negotiation framework was used in the competition. Agents were evaluated according to two categories: *individual utility* and *product of utilities*. The top four performing agents in each pool proceeded to the final for each category. Therefore, there were eight finalists for each category after the qualification round.

3 Results of the Competition

The results of the qualification and final rounds are explained in the following parts.

3.1 Results of Qualification Round

Figures 1 and 2 summarize the results of the qualification round for both Pool-1 and Pool-2, respectively. The highest four score for each category is highlighted.

According to the results of the first pool, the *Mamenchis*, *AgentKN*, *PonPokoAgent* and *SimpleAgent* qualified for the final round in the individual utility category. As far as the product of the utilities are concerned, *ParsCat2*, *taxibox*, *Mamenchis*, and *AgentKN* gained the highest product (above 0.47) and qualified for the final round.

Agent Name	Individual Utility	Product of Utilities
Farma17	0.680307731	0.38284705
ParsCat2	0.681106633	0.474018267
taxibox	0.692776131	0.487286208
Mamenchis	0.735889079	0.494775099
MadAgent	0.671866389	0.437101437
AgentKN	0.724925194	0.473772255
PonPokoAgent	0.710761165	0.448005582
SimpleAgent	0.698954726	0.437494805
Imitator	0.623136568	0.442123768

Fig. 1 Results of qualification round for Pool-1

Agent Name	Individual Utility	Product of Utilities
AgentF	0.722600455	0.39668025
GeneKing	0.683409649	0.448725557
Mosa	0.70047027	0.436250542
CaduceusDC16	0.709912278	0.421041454
ParsAgent3	0.72097358	0.435728313
TucAgent	0.675038202	0.344809219
Group3	0.623009688	0.433441349
Gin	0.696220696	0.423431405
Rubick	0.742502203	0.426399043

Fig. 2 Results of qualification round for Pool-2

In the second pool, *Agent F*, *CaduceusDC16*, *ParsAgent3* and *Rubick* agents are qualified for the final round according to the gained individual utility. In terms of the product of utilities, the agents with the highest products were *GeneKing*, *Mosa*, *ParsAgent3* and *Rubick*.

3.2 Results of Final Round

In the final round, we have separate pools for each evaluation category.

3.2.1 Individual Utility Category

In this category, we ran a tournament with the following qualified agents:

- AgentF, Nagoya Institute of Technology
- AgentKN, Nagoya Institute of Technology
- CaduceusDC16, Özyeğin University & University of Southampton
- Mamenchis, Southwest University & Maastricht University & Tianjin University
- ParsAgent3, University of Isfahan
- PonPokoAgent, Tokyo University of Agriculture and Technology
- Rubick, Özyeğin University
- SimpleAgent, Özyeğin University.

Figure 3 shows the average individual utility gained by each finalist over 1680 negotiations. Recall that each negotiating agent negotiates with all other agents five times for each negotiation scenario in Table 1. *PonPokoAgent* gained 0.75 on average and won the competition. *CaduceusDC16* and *Rubick* agents were awarded second and third place, respectively.

Three main components of the negotiation agents contribute to their overall ranking: the bidding strategy, the opponent model and the acceptance strategy [33]. In the following part, the descriptions of the agents are given.

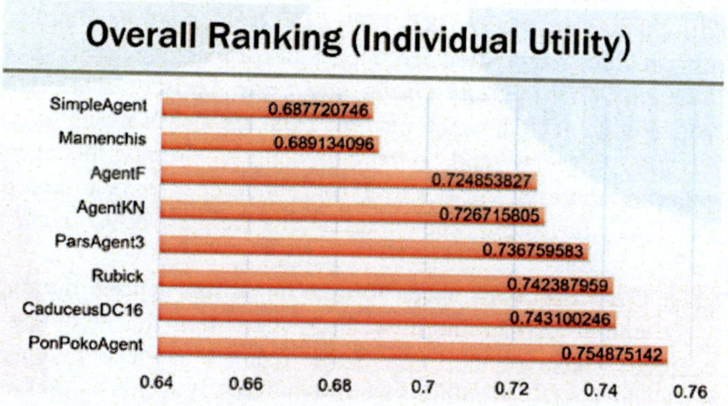

Fig. 3 Overall ranking w.r.t. individual utility

3.2.2 Winners W.r.t. Individual Utility

In this section, the detailed description of the winner strategies according to the individual utility is given.

- ***PonPokoAgent by Takaki Matsune:*** PonPokoAgent tries to propose new offers to compromise the opponents' offers as the time passes. However, it has several different bidding strategies and selects one of them randomly because it is hard for the opponents to predict our agent strategy through the previous sessions. In addition, it tries to obtain the high individual utilities just before the deadline. In the bidding strategy, it makes the list including all bids available in the domain at the beginning of the negotiation. It has two parameters to decide the highest/lowest utility of our agent's offers, and five patterns decide these parameters (time means the negotiation time elapsed [0, 1]).

  Pattern(1): $HIGHEST\_UTIL = 1.0 - 0.1 * time$
  $LOWEST\_UTIL = 1.0 - 0.1 * time - abs(sin(40 * time))$
  Pattern(2): $HIGHEST\_UTIL = 1.0 - 0.1 * time$
  $LOWEST\_UTIL = 1.0 - 0.22 * time$
  Pattern(3): $HIGHEST\_UTIL = 1.0 - 0.1 * time$
  $LOWEST\_UTIL = 1.0 - 0.15 * time - abs(sin(20 * time))$
  Pattern(4): $HIGHEST\_UTIL = 1.0 - 0.05 * time$
  $LOWEST\_UTIL = 1.0 - 0.1 * time(iftime <= 0.99)$
  $LOWEST\_UTIL = 1.0 - 0.3 * time(iftime > 0.99)$
  Pattern(5): $HIGHEST\_UTIL = 1.0 - 0.15 * time * abs(sin(20 * time))$
  $LOWEST\_UTIL = 1.0 - 0.21 * time * abs(sin(20 * time))$.

  It selects a pattern randomly at the beginning of the negotiation. It randomly selects the next offer between HIGHEST_UTIL and LOWEST_UTIL in utility from the list. In the acceptance strategy, it accepts the bid when the utility of the opponent's bid is higher than LOWEST_UTILITY. It does not use the history information from previous negotiation sessions at all for machine learning.

- ***CaduceusDC16 by Taha Gunes and Emir Arditi***: *CaduceusDC16* extends the Caduceus agent [19], which inspired from the ideas such as "*algorithm portfolio*," "*mixture of experts*" and "*genetic algorithm*." Simply, this agent asks the expert agents' opinion on whether to make a counter-offer or to accept the given offer as well as what to bid. The expert agents are the ANAC agents succeeded in the previous years.
  When CaduceusDC16 needs to take an action, it asks the expert agents what to do, and if half or more of them states that the incoming offer should be accepted, it accepts the offer. If the majority suggests making a counter-offer, then CaduceusDC16 combines bids from expert agents by picking the most favorable values of an issue by applying majority voting on each issue value. That is, the agent relies on the expert agents' bidding, opponent modeling and acceptance strategies.
  Furthermore, CaduceusDC16 aims to learn from its previous negotiations. It has an acceptance threshold and if the opponent's bids below this threshold, the agent immediately decides to make a counter-offer irrespective of expert's suggestion (i.e., reject the opponent's bid). This threshold is updated over time based on the previous negotiation experience, particularly the utility gained in the previous negotiations. For instance, if the current threshold is 0.6 and the agent had an agreement with a utility of 0.8 in its previous negotiation, then it updates the threshold to 0.8.
- ***Rubick by Okan Tunalı***: Rubick Agent is a complex time-based conceder enriched by derivations of well-studied heuristics in automated negotiation field. The main component of the agent is the target utility, which is actually lower boundary in the bid generation and in acceptance condition. If the history is not available yet, target utility is initialized as the utility of the first received bid and updated to the highest utility received from any of the opponent parties. On the other hand, if it detects a negotiation history with the same opponents, it sets the lower bound to be the highest utility ever received throughout the negotiation; thinking that the opponents will be designed in a myopic way. Both bid generation and acceptance strategies include randomness; they follow a Boulware strategy but the pace of concession is randomized. Technically, they sample from one-sided Gaussian distributions whose standard deviation increase over time, increasing the likelihood of sending bids close to lower boundary. The opponent model resolves the bids into issue evaluation values and considering their occurrence frequencies, searches for bids that holds the target utility requirements while having the most common values. That is, it employs a frequency-based opponent modeling. Finally, the model keeps a list of bids accepted by only one of the opponents in the previous negotiations, which is sorted according to the gained utility by the Rubick agent. Elements of this list are used in bid generation only if there is almost no time left. The target utility is estimated according to the following formula:
  target-utility $= \text{max} - \text{receivedUtility} + (1 - \text{max} - \text{receivedUtility}) * 1 - ((\text{current-time}^k) * |\frac{\text{Gaussian()}}{\text{constant}}|$.

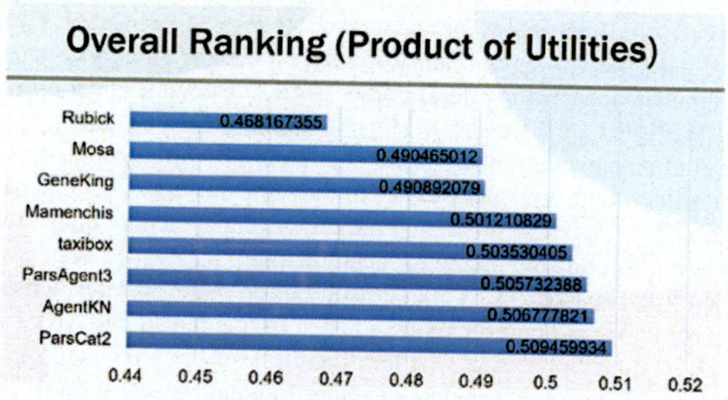

Fig. 4 Overall ranking wrt product of utility

3.2.3 Product of Utilities Category

In this category, we ran a tournament with the following qualified agents:

- AgentKN, Nagoya Institute of Technology
- GeneKing,Nagoya Institute of Technology
- Mamenchis, Southwest University & Maastricht University & Tianjin University
- Mosa, Southwest University
- ParsAgent3, University of Isfahan
- ParsCat2, University of Isfahan
- Rubick, Özyeğin University
- Tangxun, Nagoya Institute of Technology
- AgentKN, Nagoya Institute of Technology.

Figure 4 shows the average product of utilities of the agreements reached by each agent. *ParsCat2* gained 0.509 on average and won the competition where *AgentKN* and *ParsAgent3* agents took the second and third place, respectively (0.506 and 0.505). As seen from the results, the performance of these agents is very closed to each other. In the following part, the descriptions of those agents are given.

3.2.4 Winners W.r.t. Product of Utilities

In this section, the detailed description of the winner strategies according to the product utilities is given.

- ***ParsCat2 by Delaram Javdani and Maedeh Najar***: ParsCat2 mirrors changes in agent's behavior by the passage of time. Thus, it is a time-dependent agent. The main idea is that it swings the offers because it may appear flexible taking the offer as the time passes. Therefore, it keeps the offer that has good utility and would use them again.

Acceptance Strategy The acceptance strategy of ParsCat2 is comprised of multiple parts base on time: Mainly, it makes the same offer as does this agent or more than that, accepts the offer. In other cases, it calculates their utility based on time. If the utility of the agent is larger than the estimated utility, it accepts the offer from the opponent; otherwise, makes a counter-offer. At the beginning, it accepts an offer which has the utility of highest for this agent. By the passage of time, it cuts out on its expectation. On the run of time, reaching numbered seconds of the tournament, raising up the expectation again.
Bidding Strategy ParsCat2 is just going to make an offer which is good for this agent and it is good for others agents. It generates a bid, sets a threshold on the call of time and checks whether the utility of the bid calculated are between the threshold. If not, it generates another bid and repeats it on. It sometimes on the chain of search of details not finding the bid with that utility, increases the interval. In addition, it does not offer a bid with the utility of less than 0.3. If the first mover, it offers the bid with maximum utility; otherwise, it makes sure the utility of the bid to offer is bigger than the offers the other agents made. If not, it makes a bid having the same utility with that offer. It offers with swinging by the time and reduces the average of its utility of offering.

- ***AgentKN by Keita Nakamura***: AgentKN searches 10 bids that maximize self utility value while randomly shifting the initial bid. In bidding strategy, it decides the bid according to the self utility value and frequency that opponents have offered. First, it searches 10 bids that maximize self-utility. Next, it sorts 10 bids by self utility and issue values, frequency, and offers. In searching the local-optimal bids, it uses Simulated Annealing and searches 10 bids that maximize self utility value while randomly shifting the initial bid. It sorts by the following scores: (utility) $+ 0.1^{(\log_{10} \text{frequency}+1)} *$ frequency. utility is the individual utility, frequency is the sum of the number of values of issues opponents offered in the previous rounds.
  In acceptance strategy, it accepts when the utility value of the opponent's bid exceeded the threshold at that time. The threshold is decided as the following equation: $\text{threshold}(t) = 1 - (1 - \text{emax}(t)) * t^{\alpha}$ ($\text{emax}(t)$ is the estimated oppomnent's maximum utility and $\alpha(>1)$ is the parameter of conseccing to the $\text{emax}(t)$. It estimates max utility value that opponents may offer: $\text{emax}(t) = \mu(t) + (1 - \mu(t))d(t)$
  $d(t) = \frac{\sqrt{3}\sigma(t)}{\sqrt{\mu(t)(1-\mu(t))}}$
  ($\mu(t)$: Average of utility values that the opponent has offered, $d(t)$: Estimated width of the bid range of the opponent, which uses the concession strategy, $\sigma(t)$: Standard deviation of utility value that the opponent has offered).
- ***ParsAgent3 by Zahra Khosravimehr and Faria Nassiri-Mofakham***: ParsAgent3 proposes a negotiation strategy in the multilateral negotiation domain. It is a concession-based negotiation strategy. It begins with a concession. If its opponents concede, it continues with more concessions, otherwise, it changes its strategy and behaves hardheaded. To this end, ParsAgent3 first calculates its temporary target utility. It uses this value for using in bidding and acceptance strategies according

to the following equations:

$$u(t) = P_{\min} + (P_{\max} - P_{\min})(1 - F(t))$$

$$F(t) = k + (1 - k)t^{1/\beta}$$

In these equations, t is time, β indicates the agent's concedence speed during the negotiation process. The variables, $P_{\min}$ and $P_{\max}$ are respectively the minimum and maximum values of the function $F(t)$, so that $P_{\min}, P_{\max}, k \in [0, 1]$.
ParsAgent3 clusters every 100 bids received from each individual opponent and estimates the temporary reservation value of each opponent [24]. This value is the minimum acceptable utility for the opponent at the given time. The agent then sets its reservation value according to the current estimated reservation values of all the opponents. If the opponents concede, the average of their reservation values is employed. Otherwise, the maximum value among the estimated reservation values of all the opponents is chosen as ParsAgent3 reservation value.
It calculates the difference between maximum and minimum utilities of the bids received from each opponent in the previous session using the history information. Then, if this value is less than 0.3, ParsAgent3 categorizes the opponents as hardheaded and so the agent will not concede in the current session.

4 Related Work

The automated negotiating agent competition follows in the footsteps of a series of successful competitions that aim to advance the state-of-the-art in artificial intelligence (other examples include the Annual Computer Poker Competition and the various Trading Agent Competitions, the Alimama International Advertising Algorithm Competition and Angry Birds Competition). Over the years, the competition has addressed various topics: varying the number of negotiators, the complexity of the negotiation domains, repeated negotiations with the same set of opponents, negotiations in special domains and negotiating against humans.

Every year, new features have been incorporated into the competition environment to increase realism and to encourage the development of flexible and practical negotiation agents [6]. The first installment of ANAC was in 2010 and started as a jointly organized competition between Delft University of Technology and Bar-Ilan University [12]. Since the first edition, the local organization has rotated between various international institutions. In 2011, ANAC was organized by the 2010 winner, the Nagoya Institute of Technology [14]. Next editions were organized by University of Southampton (2012) [36], Ben Gurion University of the Negev (2013) [18], Nagoya Institute of Technology and Tokyo University of Agriculture and Technology (2014) [17], and Delft University of Technology (2015) [7].This paper describes the competitions from the next subsequent years. Every installment has seen a growing number of participants: ANAC 2010 started with seven teams from five different institutes, while in 2017, it featured 18 teams from nine institutes.

By analyzing the results of ANAC, the stream of the strategies of ANAC and important factors for developing the competition have been shown. Baarslag et al. [34] presented an in-depth analysis of ANAC 2011. This paper mainly analyzes the different strategies using classifications of agents with respect to their concession behavior against a set of standard benchmark strategies and empirical game theory (EGT) to investigate the robustness of the strategies. It also shows that the most adaptive negotiation strategies, while robust across different opponents, are not necessarily the ones that win the competition. Furthermore, our EGT analysis highlights the importance of considering metrics.

Kawaguchi et al. [30] proposed a strategy for compromising the estimated maximum value based on estimated maximum utility. These papers have been important contributions for bilateral multi-issue closed negotiation; however, they don't deal with multi-time negotiation with learning and reusing the past negotiation sessions. This agent is the winner of ANAC2010. After that, Fujita [15, 16] proposed the compromising strategy with adjusting the speed of making agreements using the Conflict Mode and focused on multi-time negotiations. However, these strategies only focused on the linear utility functions.

In addition, some studies have focused on the divided parts of negotiating strategies in the alternating offering protocol: proposals, responses, and opponent modeling. Effective strategies can be achieved by combinations of top agents' strategies in the competitions depending on the opponent's strategies and negotiation environments. Many of the sophisticated agent strategies that currently exist are comprised of a fixed set of modules. Therefore, the studies for proposing the negotiation strategies focusing on the modules are important and influential [32, 37].

5 Conclusion & Future Work

This paper describes the ANAC 2017 Repeated Multilateral Negotiation Competition League. Learning from previous negotiations and enhancing the current negotiation strategy based on those experiences are the main challenges introduced in this competition. According to the results of the qualification and final round, it can be observed that employing different negotiation strategies against the same opponents in a repeated negotiation setup may let the agent gain higher utility on average (e.g., PonPoko Agent). Although the participants introduced new ideas regarding the underlying challenges, there is still some room for improving and designing more sophisticated agents. For this reason, organizers have introduced the same challenge in ANAC 2018 to enrich the benchmark in repeated multilateral negotiation research.

There are a number of new challenges planned for future iterations of the competition. For 2019, we aim to introduce the challenge of *negotiating under preference uncertainty* [13, 27, 31]. Instead of a fully specified utility function, agents in 2019 will be tasked with performing a successful negotiation when only partial preferences is available in the form of ordinal preference comparisons (i.e., "outcome x is preferred over outcome y"). The overarching goal is to design negotiating agents that will eventually be able to represent users in a negotiation, even when limited preference elicitation can be performed (e.g., due to user bother).

Acknowledgements This work is part of the Veni research program with project number 639.021.751, which is financed by the Netherlands Organisation for Scientific Research (NWO) and by the ITEA M2MGrids Project, grant number ITEA141011.

References

1. R. Aydoğan, D. Festen, K.V. Hindriks, C.M. Jonker, Alternating offers protocols for multilateral negotiation, in *Modern Approaches to Agent-based Complex Automated Negotiation*, ed. by K. Fujita, Q. Bai, T. Ito, M. Zhang, F. Ren, R. Aydoğan, R. Hadfi (Springer International Publishing, Berlin, 2017), pp. 153–167
2. R. Aydoğan, K.V. Hindriks, C.M. Jonker, Multilateral mediated negotiation protocols with feedback, in *Novel Insights in Agent-based Complex Automated Negotiation* (Springer, Berlin, 2014), pp. 43–59
3. R. Aydoğan, P. Yolum, The effect of preference representation on learning preferences in negotiation, in *New Trends in Agent-Based Complex Automated Negotiations*, Vol. 383, Studies in Computational Intelligence, ed. by T. Ito, M. Zhang, V. Robu, S. Fatima, T. Matsuo (Springer, Berlin, 2012), pp. 3–20
4. R. Aydoğan, T. Baarslag, K.V. Hindriks, C.M. Jonker, P. Yolum, Heuristics for using cp-nets in utility-based negotiation without knowing utilities. Knowl. Inform. Syst. **45**(2), 357–388 (2014)
5. T. Baarslag, Accepting optimally in automated negotiation with incomplete information, in *Proceedings of the 25th Benelux Conference on Artificial Intelligence* (2013)
6. T. Baarslag, *What to bid and when to stop*. Dissertation, Delft University of Technology, Sep 2014
7. T. Baarslag, R. Aydoğan, K.V. Hindriks, K. Fuijita, T. Ito, C.M. Jonker, The automated negotiating agents competition, 2010–2015. AI Mag. **36**(4), 115–118 (2015)
8. T. Baarslag, A.S.Y. Dirkzwager, K.V. Hindriks, C.M. Jonker, The significance of bidding, accepting and opponent modeling in automated negotiation. In *21st European Conference on Artificial Intelligence*, vol. 263 of *Frontiers in Artificial Intelligence and Applications* (2014), pp. 27–32
9. T. Baarslag, E.H. Gerding, Optimal incremental preference elicitation during negotiation, in *Proceedings of the Twenty-fourth International Joint Conference on Artificial Intelligence (IJCAI'15)*. AAAI Press (2015), pp. 3–9
10. T. Baarslag, E.H Gerding, R. Aydoğan, M.C. Schraefel, Optimal negotiation decision functions in time-sensitive domains, in *2015 IEEE/WIC/ACM International Conference on Web Intelligence and Intelligent Agent Technology (WI-IAT)*, vol. 2. IEEE (2015), pp. 190–197
11. T. Baarslag, M.J.C. Hendrikx, K.V. Hindriks, C.M. Jonker, Learning about the opponent in automated bilateral negotiation: a comprehensive survey of opponent modeling techniques. *Autonomous Agents and Multi-Agent Systems* (2015), pp. 1–50
12. T. Baarslag, K. Hindriks, C.M. Jonker, S. Kraus, R. Lin, The first automated negotiating agents competition (ANAC 2010), in *New Trends in Agent-based Complex Automated Negotiations, Series of Studies in Computational Intelligence*, ed. by T. Ito, M. Zhang, V. Robu, S. Fatima, T. Matsuo (Springer, Berlin, 2012) , pp. 113–135
13. T. Baarslag, M. Kaisers, The value of information in automated negotiation: a decision model for eliciting user preferences. In *Proceedings of the 16th Conference on Autonomous Agents and*

MultiAgent Systems (AAMAS '17), Richland, SC. International Foundation for Autonomous Agents and Multiagent Systems (2017), pp. 391–400
14. T. Baarslag, K. Fujita, E.H. Gerding, K. Hindriks, T. Ito, N.R. Jennings, C. Jonker, S. Kraus, R. Lin, V. Robu, C.R. Williams, Evaluating practical negotiating agents: results and analysis of the 2011 international competition. Artif. Intell. **198**, 73–103 (2013). May
15. K. Fujita, Automated strategy adaptation for multi-times bilateral closed negotiations, in *Proceedings of the 13th International Conference on Autonomous Agents and Multiagent Systems (AAMAS 2014)* (2014), pp. 1509–1510
16. K. Fujita, Compromising adjustment strategy based on tki conflict mode for multi-times bilateral closed negotiations. Comput. Intell. (2017)
17. K. Fujita, R. Aydoğan, T. Baarslag, T. Ito, C. Jonker, The fifth automated negotiating agents competition (ANAC 2014), in *Recent Advances in Agent-based Complex Automated Negotiation*, ed. by N. Fukuta, T. Ito, M. Zhang, K. Fujita, V. Robu (Springer International Publishing, Cham, 2016), pp. 211–224
18. K. Gal, T. Ito, C. Jonker, S. Kraus, K. Hindriks, R. Lin, T. Baarslag, in *The Forth International Automated Negotiating Agents Competition (ANAC2013)* (2013)
19. T.D. Güneş, E. Arditi, R. Aydoğan, Collective voice of experts in multilateral negotiation, in *Principles and Practice of Multi-Agent Systems (PRIMA 2017)*. Lecture Notes in Computer Science, vol 10621 (Springer, Cham, 2017), pp. 450–458
20. K. Hara, T. Ito, A mediation mechanism for automated negotiating agents whose utility changes over time, in *Twenty-Seventh AAAI Conference on Artificial Intelligence* (2013)
21. I. Marsa-Maestre, M. Klein, C.M. Jonker, R. Aydoğan. From problems to protocols: Towards a negotiation handbook. Decis. Support Sys.**60**, 39–54 (2014)
22. C.M. Jonker, R. Aydoğan, T. Baarslag, K. Fujita, T. Ito, K. Hindriks, Automated negotiating agents competition (anac), in *Proceedings of the Thirty-First AAAI Conference on Artificial Intelligence (AAAI-17)*. AAAI Press (2017), pp. 5070–5072
23. S. Kawaguchi, K. Fujita, T. Ito, Compromising strategy based on estimated maximum utility for automated negotiation agents competition (anac-10). Mod. Approach. Appl. Intell. Lect. Notes Comput. Sci. **6704**, 501–510 (2011)
24. Z. Khosravimehr, F. Nassiri-Mofakham. Pars agent: hybrid time-dependent, random and frequency-based bidding and acceptance strategies in multilateral negotiations, in *Modern Approaches to Agent-based Complex Automated Negotiation*, 1st ed., ed. by K. Fujita, Q. Bai, T. Ito, M. Zhang, F. Ren, R. Aydogan, R. Hadfi (Springer Publishing Company, Incorporated, 2017), pp. 175–183
25. M. Klein, P. Faratin, H. Sayama, Y. Bar-Yam, Negotiating complex contracts. Group Decis. Negot. **12**(2), 58–73 (2003)
26. R. Lin, S. Kraus, T. Baarslag, D. Tykhonov, K. Hindriks, C.M. Jonker, Genius: an integrated environment for supporting the design of generic automated negotiators. Comput. Intell. **30**(1), 48–70 (2014)
27. X. Luo, N.R. Jennings, N. Shadbolt, Acquiring user strategies and preferences for negotiating agents: a default then adjust method. Int. J. Hum. Comput. Stud. **64**(4), 304–321 (2006)
28. I. Marsa-Maestre, M. Klein, C.M. Jonker, R. Aydoğan, From problems to protocols: yowards a negotiation handbook, in *Decision Support Systems* (2014), pp. 39–54
29. V. Sánchez-Anguix, S. Valero, V. Julián, V. Botti, A. García-Fornes, Evolutionary-aided negotiation model for bilateral bargaining in ambient intelligence domains with complex utility functions. Inform. Sci. **222**, 25–46 (2013)
30. Shogo Kawaguchi, Katsuhide Fujita, and Takayuki Ito. Compromising strategy based on estimated maximum utility for automated negotiation agents competition (anac-10). In *24th International Conference on Industrial Engineering and Other Applications of Applied Intelligent Systems (IEA/AIE d2011)*, (2011) pp. 501–510
31. T. Baarslag, E.H. Gerding, Optimal incremental preference elicitation during negotiation. In *Proceedings of the Twenty-fourth International Joint Conference on Artificial Intelligence IJCAI-15*. AAAI Press, (2015) pp. 3–9

32. T. Baarslag, K.V. Hindriks, Accepting optimally in automated negotiation with incomplete information. In *Proceedings of the 12th International conference on Autonomous agents and multi-agent systems (AAMAS 2013)*, (2013) pp. 715–722
33. T. Baarslag, S.Y. Alexander, Dirkzwager, K.V. Hindriks, M.J. Catholijn, The significance of bidding, accepting and opponent modeling in automated negotiation. In *21st European Conference on Artificial Intelligence*, vol. 263 of Frontiers in Artificial Intelligence and Applications (2014) pp. 27–32
34. Tim Baarslag, Katsuhide Fujita, Enrico Gerding, Koen Hindriks, Takayuki Ito, Nick R. Jennings, Catholijn Jonker, Sarit Kraus, Raz Lin, Valentin Robu, and Colin Williams. Evaluating practical negotiating agents: Results and analysis of the 2011 international competition. Artif. Intell. **198** (2013) pp. 73–103
35. Víctor Sánchez-Anguix, Soledad Valero, Vicente Julián, Vicente Botti, and Ana García-Fornes. Evolutionary-aided negotiation model for bilateral bargaining in ambient intelligence domains with complex utility functions. Inform. Sci. **222**, 25–46 (2013)
36. R.C. Williams, V. Robu, E.H. Gerding, N.R. Jennings, An overview of the results and insights from the third automated negotiating agents competition (ANAC 2012), in *Novel Insights in Agent-based Complex Automated Negotiation*, vol. 535, Studies in Computational Intelligence, ed. by I. Marsa-Maestre, M.A. Lopez-Carmona, T. Ito, M. Zhang, Q. Bai, K. Fujita (Springer, Japan, 2014), pp. 151–162
37. F. Zafari, F.N. Mofakham, POPPONENT: highly accurate, individually and socially efficient opponent preference model in bilateral multi issue negotiations, in *Proceedings of the Twenty-Sixth International Joint Conference on Artificial Intelligence (IJCAI 2017)* (2017), pp. 5100–5104

Cooperativeness Measure Based on the Hypervolume Indicator and Matching Method for Concurrent Negotiations

Ryohei Kawata and Katsuhide Fujita

Abstract Multi-agent systems involve concurrent negotiation problems wherein many agents negotiate concurrently; thus, such problems have gained research attention as one of the most important challenges. Considering bilateral negotiation with all opponents is unrealistic when the negotiation problems involve many agents. In this situation, agents attempt to negotiate with cooperative opponents to improve the negotiation result. However, discovering cooperative opponents is difficult because there is no reliable measure to evaluate cooperativeness. In this paper, we propose a cooperativeness measure based on the hypervolume indicator (CHI) to evaluate the cooperativeness among agents. CHI is the hypervolume of the negotiation space, which is Pareto dominated by the set of all the bids. Furthermore, we propose a matching method based on the cooperativeness measure for concurrent negotiations. The proposed method finds the matching by maximizing the sum of cooperativeness of each pair. In the experiments, we evaluate CHI and our matching method by analyzing the negotiation results of three state-of-the-art agents in ANAC 2016. The experimental results demonstrate the following: (1) there is an extremely strong correlation between CHI and social welfare and (2) total social welfare of a matching by our method is quite close to the theoretical optimal.

1 Introduction

Automated negotiation is playing an important role in multi-agent systems [8, 12, 13]. In a multi-agent system, cooperation among the agents increases the social welfare [21]. However, it is difficult for agents to cooperate when each agent has its own different objective and utility function. Automated negotiation can resolve conflicts and enable the agents to cooperate with each other. In addition, developing auto-

R. Kawata (✉) · K. Fujita
Tokyo University of Agriculture and Technology, Koganei, Tokyo, Japan
e-mail: kawata@katfuji.lab.tuat.ac.jp

K. Fujita
e-mail: katfuji@cc.tuat.ac.jp

T. Ito et al. (eds.), *Advances in Automated Negotiations*, Studies in Computational Intelligence 905, https://doi.org/10.1007/978-981-15-5869-6_8

mated negotiation is expected to contribute to support human negotiations and lead to the realization of a decision support system. The International Automated Negotiating Agent Competition (ANAC) has been held annually since the first competition in 2010 [1, 2, 4]. In ANAC, many state-of-the-art agents have been proposed and it contributes to the development of automated negotiation. Moreover, through the development of GENIUS [15], which is a negotiation platform used in ANAC, and Negowiki [16, 17], the negotiation protocols and scenarios have been standardized.

Multi-issue multilateral negotiation is an important class of automated negotiation because its settings are almost identical to real-life negotiations. Multi-issue negotiation comprises multiple issues wherein each issue has some options. The stacked alternative offers protocol (SAOP) is one of the most frequently used protocols in multilateral negotiations [3]. In SAOP, agents propose or accept in turn. When all agents accept the same offer, the agreement is achieved.

Concurrent negotiation, in which many agents negotiate concurrently, is almost identical to real negotiation problems encountered in supply chain management and smart grids. Therefore, concurrent negotiation is one of the most important challenges for automated negotiation. In concurrent negotiation, it is unrealistic to negotiate bilaterally with all opponents when the negotiation involves many agents. Negotiating with cooperative opponents is effective to improve negotiation results. Therefore, a reliable measure must be implemented to evaluate the cooperativeness between agents for finding cooperative pairs. For example, metric of opposition level (MOL) is the measure that evaluates the difficulty in reaching an agreement [17, 22]. MOL evaluates domains based on the fairness of the utilities of agents. Since MOL focuses only on fairness, it regards some domains as the cooperative domains in spite of the expected social welfare being small. To solve this problem, we propose the cooperativeness measure based on the hypervolume indicator (CHI) as a cooperativeness measure. CHI evaluates the cooperativeness based on the hypervolume of the negotiation space, which is based on Pareto dominants as a set of all the bids of a domain. This hypervolume is called the hypervolume indicator: frequently applied to evaluate the results of evolutionary multi-objective optimization algorithms [23]. In this paper, CHI is applied to matching. However, CHI can also be applied to domain evaluation for negotiation analysis or competitions. Furthermore, we propose a matching method based on the cooperativeness measure for concurrent negotiations. The proposed method generates the matching by maximizing the sum of the cooperativeness of each pair.

We experimentally evaluate CHI and the proposed matching method by analyzing the negotiation results of the top three state-of-the-art agents in ANAC 2016. To evaluate the cooperativeness measures, we compared the correlation coefficients between the social welfare of these negotiations and each cooperativeness measure. The results demonstrate that CHI has an extremely strong correlation to the social welfare of the negotiation results. Moreover, we compare the total social welfare between the matching generated using the proposed method against the theoretical optimal solution. The experimental results show that the proposed method can find matchings that are similar to the theoretical optimal solutions.

This paper is organized as follows. First, we describe the definition of the negotiation environments that comprise multi-issue negotiation, concurrent negotiation, and SAOP. Next, we propose cooperativeness measures to evaluate the relations between agents and the matching method for concurrent negotiations. Then, we demonstrate experimental results. Finally, we discuss the related works in automated negotiations and describe the conclusions.

2 Negotiation Environments

2.1 Multi-issue Negotiation

This paper considers the multi-issue negotiation problem. Negotiation domains define negotiation spaces. In multi-issue negotiation problems, negotiation domain D has a set of issues $I = \{i_1, i_2, \ldots, i_n\}$. Each issue $i \in I$ has a set of values $V^i = \{v_1^i, v_2^i, \ldots, v_{m_i}^i\}$. Here, n is the number of issues of D and m_i is the number of values of i.

A bid is defined as a set of $v^i \in V^i$ for each $i \in I$, and a bid b is expressed as $b = (v^{i_1}, v^{i_2}, \ldots, v^{i_n})$. Agents select an action from proposing, accepting, or rejecting bids in the negotiations. A set B that comprises all bids is

$$B = \{b = (v^{i_1}, v^{i_2}, \ldots, v^{i_n}) \mid v^i \in V^i, i \in I\}.$$

Each agent has its own utility function $U(b)$, including weight w^i for each $i \in I$ $\left(\sum_{i \in I} w^i = 1\right)$ and evaluation value $eval(v^i)$ for each $v^i \in V^i, i \in I$. The utility of $b \in B$ is

$$U(b) = \sum_{v^i \in b} w^i \cdot \frac{\mathrm{eval}(v^i)}{\max_{v \in V^i} \mathrm{eval}(v)}.$$

Each agent has its unique set of weights of I and evaluation values of V to express its preference.

Negotiations have a timeline t, which is normalized to the range $[0, 1]$. $t = 0$ represents the time of starting a negotiation, and $t = 1$ means a negotiation has passed the deadline. An agent has a discount factor $\delta \in [0, 1]$ that decreases the agent's utility depending on t [7].U_a is the utility func Discounted utility U_{DF} is

$$U_{\mathrm{DF}}(b, t) = U(b) \cdot \delta^t.$$

$\delta = 1$ means the utility is independent of time t, and $\delta < 1$ show that the utility decreases with increasing t. The discount factors are one motivation for agents to quickly reach an agreement.

An agent has the reservation value u_{RV}, which is applied to the utility of the agent when it fails to make an agreement. Considering the discount factor, the utility in failed negotiations is

$$U_{\mathrm{RV}}(t) = u_{\mathrm{RV}} \cdot \delta^{t}.$$

$u_{\mathrm{RV}} > 0$ means that an agent can obtain utility when the negotiations failed.

2.2 Concurrent Negotiation

In this paper, the concurrent negotiation problem is defined as a situation wherein many agents are involved bilateral negotiations with any opponent in concurrently. Let A be a set of agents. A set of all negotiable pairs P is given as $P = \{p \subseteq A \mid |p| = 2\}$. Pair $p \in P$ can negotiate and agents can join more than two negotiations concurrently. In a concurrent negotiation, it is unrealistic to negotiate bilaterally with all opponents under negotiations involving too many agents. Therefore, finding a matching between agents which can cooperate with each other assists in reaching a better agreement in concurrent negotiation problems.

In this paper, a matching $m \in M$ in concurrent negotiation problems is expressed as a set of negotiable pairs, where M is a set of all matchings and it is defined as follows:

$$M = \{m \subseteq P \mid p_1, p_2 \in m, p_1 \cap p_2 = \phi\}.$$

Let $\mathrm{eval}_P(p)$ be an evaluation value of $p \in P$. An evaluation value $\mathrm{eval}_M(m)$ of matching $m \in M$ is

$$\mathrm{eval}_M(m) = \sum_{p \in m} \mathrm{eval}_P(p).$$

In this paper, we use social welfare of p as $\mathrm{eval}_P(p)$.

The objective function of finding the matchings in concurrent negotiation problems is

$$\arg\max_{m \in M} = \mathrm{eval}_M(m).$$

By considering A as the set of nodes and cooperativeness of p as weights of the edges, finding matchings in concurrent negotiation problems can be considered as the maximum weighted matching problem.

2.3 Stacked Alternative Offers Protocol

In this paper, the negotiation protocol mainly uses SAOP [3], which is an extension of a bilateral alternative offers protocol [19] for multilateral negotiations. The flow

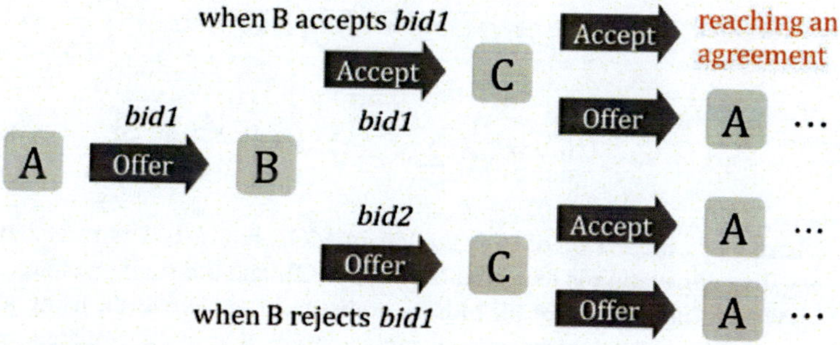

Fig. 1 SAOP flow chart

chart of SAOP among agents A, B, and C is shown in Fig. 1. In SAOP, agents take an action in turns and the order of agents is immutable. Each agent can take the following action.

- *Offer*: Rejecting the previous bid and proposing a new one to the next agent.
- *Accept*: Accepting the previous bid
- *EndNegotiation*: Terminating the negotiation without agreement.

The agents continue this process until any one condition among the following conditions is satisfied.

- A bid is accepted by all agents, and they can reach an agreement.
- The negotiation deadline passed before reaching an agreement.
- An agent terminates the negotiation by selecting *EndNegotiation*.

In reaching an agreement, each agent obtains utility $U(b, t)$, where b is the bid accepted by all agents and t is the timeline of reaching an agreement. When *EndNegotiation* is selected at time t, the utility of each agent is $U_{\mathrm{RV}}(t)$. The utility of each agent is $U_{\mathrm{RV}}(t = 1)$ when the deadline has been passed before reaching an agreement. In SAOP, the deadline is defined as the real-time or the number of agent's actions, which are called as rounds.

3 Cooperativeness Measures and Matching Method

3.1 Existing Measure: Metric of Opposition Level

The metric of opposition level (MOL) is a measure to evaluate the difficulty in reaching an agreement [17, 22]. MOL evaluates domains based on the fairness of the distribution of all the bids. When the difference of the agents' utilities of a bid is small, MOL evaluates the bid as fair. Let A be a set of agents compose of a negotiation, U_a be the utility function of $a \in A$. MOL is defined as

$$\mathrm{MOL}(A, \Omega) = z \sum_{\omega \in \Omega} \sum_{a \in A} \left(\overline{U(\omega)} - U_a(\omega) \right)^2,$$

$$\overline{U(\omega)} = \frac{1}{|A|} \sum_{a \in A} U_a(\omega),\ z = \frac{|A|}{|A| - 1} \cdot \frac{1}{|\Omega|},$$

where Ω is the set of bids evaluated in MOL and z is a normalization factor. A negative correlation is expected between MOL and the social welfare.

We can use the set of all bids in domains in Ω. However, using all the bids in a domain is inefficient because it includes bids that are not related to agreements. Therefore, MOL can improve the correlations with negotiation results by using either a set of all the Pareto optimal bids or a set of all bids in which the utility is more than the reservation value in Ω.

Figure 2 shows a concept of fair bids and unfair bids in MOL. For the fair bids, the difference between the agents' utilities is small. Therefore, all the agents are likely to agree to the fair bid. In contrast, for the unfair bids, the difference between the agents' utilities is large. Therefore, the unfair bids are likely to be rejected by the agents.

MOL evaluates the domains by focusing only on fairness and not on the social welfare. Therefore, in some situations, MOL regards a domain as cooperative in spite of a small expected social welfare.

3.2 *Existing Measure: Correlation Metric*

Correlation metric evaluates the difficulty of a negotiation based on the correlation between utility functions for different agents [17]. In this paper, we define *correlation*

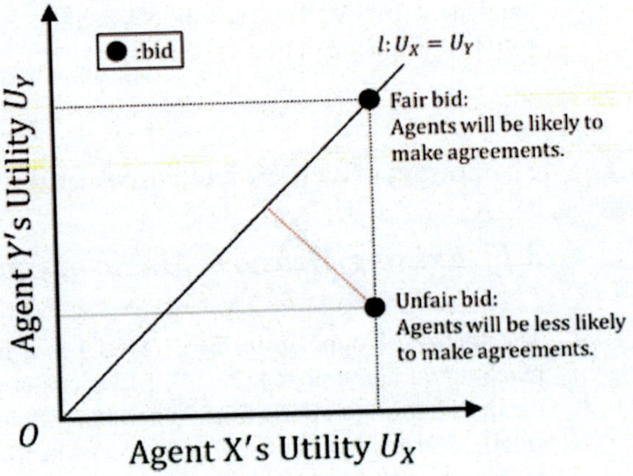

Fig. 2 Concept of fair bids and unfair bids in MOL

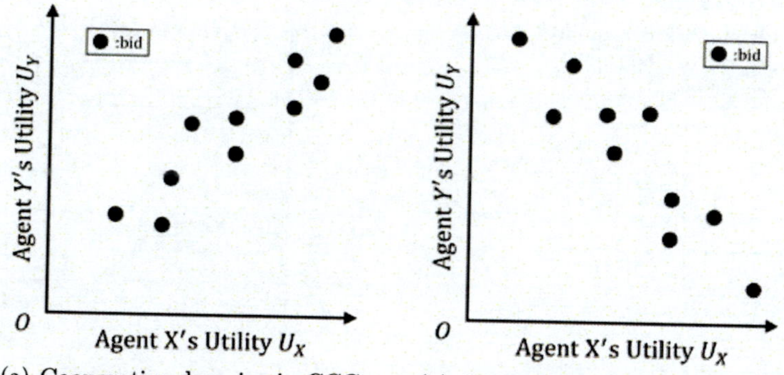

(a) Cooperative domains in CCC (b) Competitive domains in CCC

Fig. 3 Typical utility diagrams of cooperative domains and competitive domains in CCC

metric as a correlation coefficient between utility functions, which normalized to the interval [0, 1], and we call it the cooperativeness measure based on the correlation coefficient (CCC). CCC cannot be applied to a multilateral negotiation because of using correlation coefficient. Cases wherein agents X, Y negotiate the set of all bids B, U_X is the utility function of X and U_Y is the utility function of Y; the value of CCC is

$$CCC(X, Y, B) = \frac{CC(\{(U_X(b), U_Y(b)) \mid b \in B\}) + 1}{2},$$

where CC is a function of a correlation coefficient. A positive correlation is expected between CCC and the social welfare.

Figure 3 shows the typical utility diagrams of cooperative domains and competitive domains in CCC. CCC assumes that the correlation between the utilities of agents is positive in cooperative domains. In contrast, CCC assumes the correlation of utilities is negative in competitive domains.

3.3 Cooperativeness Measure Based on the Hypervolume Indicator

We propose the cooperativeness measure based on the hypervolume indicator (CHI) to evaluate the cooperative relation between agents. CHI is defined as the hypervolume of the negotiation space, which is Pareto dominated by the set of all bids of a domain in a utility diagram. CHI can be applied to any N-lateral negotiations because of based on the hypervolume indicator.

The hypervolume indicator [23] is one of the most frequently used indicators applied to the evaluation of results from evolutionary multi-objective optimization

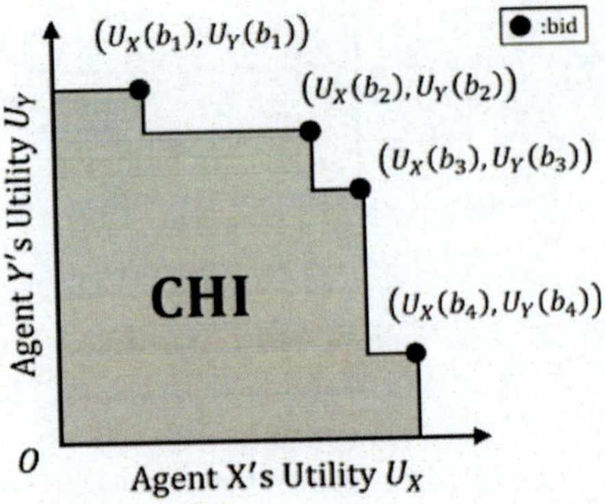

Fig. 4 CHI in the utility diagram of a bilateral negotiation

algorithms. The hypervolume indicator is the hypervolume of the space dominated by a set of solutions and bounded by a reference point. Given a set of solutions S and a reference point r, let $HI(S, r)$ be the hypervolume indicator defined by S and r, $\mathrm{HI}(S, r)$ is

$$\mathrm{HI}(S, r) = H\left(\bigcup_{s \in S} \text{hyperrectangle}[s, r]\right),$$

where $H(h)$ is the hypervolume of h and $hyperrectangle[s, r]$ is the hyperrectangle defined by s and r.

When a set of agents A negotiate about a set of all bids B, let O be the origin. The value of CHI is

$$\mathrm{CHI}(A, B) = H\left(\bigcup_{b \in B} \text{hyperrectangle}[b, O]\right),$$

where hyperrectangle$[b, O]$ is the hyperrectangle defined by b and O in the utility diagram among A. A positive correlation is expected between CHI and the social welfare.

Figure 4 shows CHI in a bilateral negotiation with agents X, Y. B^* is a set of all Pareto optimal bids, U_X is the utility function of X and U_Y is the utility function of Y. The negotiation space with Pareto dominated by $b^* \in B^*$ is the rectangle defined by point $(U_X(b^*), U_Y(b^*))$ and O. The value of CHI is the area of the union of all the rectangles.

Algorithm 1: Proposed matching method

```
Data: A: Set of agents in the negotiations, C: Cooperativeness measure used for finding
      matchings.
begin
    for a ∈ A do
        preference[a] ← get_preference(a)
                                        // collect preference of agent a
    for a1 ∈ A do
        for a2 ∈ A \ {a1} do
            coop[(a1, a2)] ← calculate_cooperativeness(a1, a2, C)
                // calculate the cooperativeness of pair (a1, a2) by C
    if is_positive_correlation(C) then
            // Does C expected to have a positive correlation with
        the social welfare? (Does C equal CHI or CCC?)
        m* ← generate_matching_positive(A, coop)
                // generate a matching by maximizing the sum of coop
    else
        m* ← generate_matching_negative(A, coop)
                // generate a matching by minimizing the sum of coop
    return m*
```

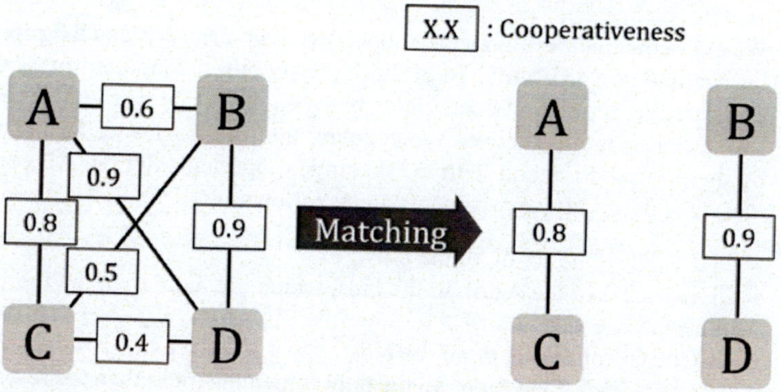

Fig. 5 Example of matching by proposed method

3.4 *Matching Method Based on the Cooperativeness Measure for Concurrent Negotiations*

We propose a matching method based on the cooperativeness measure for concurrent negotiations. In the proposed method, the mediator tries to find the cooperative pairs that agents can reach a better agreement using cooperativeness measures. Figure 5 shows an example of finding a matching using our method in a concurrent negotiation by four agents: A, B, C, and D. The pseudocode of our method is shown in Algorithm

1. First, the mediator collects their preference information of all the agents of A in domain D, where A is the set of all the agents. Second, the mediator calculates the cooperativeness of each negotiable pair $p \in P$, where $P = \{p \subseteq A \mid |p| = 2\}$ is the set of all negotiable pairs. Finally, the mediator tries to find a matching m^* by maximizing the sum of cooperativeness of each $p \in m^*$, where $m^* \in M$ and $M = \{m \subseteq P \mid p_1, p_2 \in m, p_1 \cap p_2 = \phi\}$ is the set of all the matchings. When the cooperativeness measure has a negative correlation with the social welfare like MOL, the mediator find a matching by minimizing the sum of cooperativeness of each pair. By considering a set of agents A as a set of nodes and $eval_P$ as weights of edges, to find matchings in concurrent negotiation problems can be considered as the maximum weighted matching problem. Therefore, the mediator generates a matching m^* by resolving maximum weighted matching problems. In this paper, the mediator uses blossom algorithm and primal-dual algorithm to solve maximum weighted matching problems [9].

4 Experiments

4.1 Experimental Settings

We experimentally evaluate the cooperativeness measures and the proposed matching method using GENIUS. To evaluate the proposed cooperativeness measures, we compare the correlation coefficients between the social welfare and each cooperativeness measure. Moreover, we compare the total social welfare of the matching by the proposed method with the theoretical optimal solution. All cooperativeness measures used in the experiments are as follows:

- CHI: CHI of the set of all the bids.
- CHI_{RV}: CHI of the set of all the bids, which are more than the reservation values of all the agents.
- CCC: CCC of the set of all the bids.
- CCC_{RV}: CCC of the set of all the bids, which are more than the reservation values of all the agents.
- MOL_{ALL}: The MOL wherein Ω is the set of all the bids.
- MOL_{RV}: The MOL wherein Ω is the set of all the bids which are more than the reservation values of all the agents.
- MOL_{PO}: The MOL wherein Ω is the set of all the Pareto optimal bids.
- MOL_{RVPO} The MOL wherein Ω is the set of all the Pareto optimal bids with reservation values of all of the agents.

We use several domains of GENIUS and the new domains generated by us to evaluate the cooperativeness in various situations. Tables 1 and 2 show the list of domains and their characteristics used herein. The issues are independent in all domains. Each domain of the GENIUS has four to eight profiles. A profile comprises

Table 1 GENIUS domains used in the experiments

Domain	Num of issues	Num of values	Reservation value	Discount factor
domain1	1	5	0.5	1.0
domain2	2	5	0 or 0.5	0.2 or 1.0
domain4	4	4 or 5	0	0.5
domain8	8	3	0	1.0
domain16	16	2	0.7	0.4
party_domain (profile 1–8)	6	3 or 4	0	1.0

Table 2 Our domains used in the experiment

Domain	No. of issues	No. of values	$\text{eval}(v^i)$
domain_3a	3	5	$\text{rand}_{[0,1)}$
domain_3b	3	5	$(\text{rand}_{[0,1)})^2$
domain_3c	3	5	$1-(\text{rand}_{[0,1)})^2$
domain_5a	5	5	$\text{rand}_{[0,1)}$
domain_5b	5	5	$(\text{rand}_{[0,1)})^2$
domain_5c	5	5	$1-(\text{rand}_{[0,1)})^2$
domain_7a	7	5	$\text{rand}_{[0,1)}$
domain_7b	7	5	$(\text{rand}_{[0,1)})^2$
domain_7c	7	5	$1-(\text{rand}_{[0,1)})^2$

the weights of $i \in I$, the evaluation value of $v^i \in V^i$, the reservation value, and the discount factor to define the preference of the agent. Each of our domains has 20 profiles. The reservation value of all our domains is 0.5 and a discount factor is 1.0. In our domains, $\text{eval}(v^i)$ is decided randomly. When $\text{eval}(v^i) = \text{rand}_{[0,1)}$, the bids are distributed at the center of the utility diagrams, as shown in Fig. 6, where $\text{rand}_{[0,1)}$ is a uniform random number of interval [0, 1). When $eval(v^i) = (\text{rand}_{[0,1)})^2$, the agents have a lot of bids with a low utility, and it as shown in Fig. 7. When $\text{eval}(v^i) = 1 - (\text{rand}_{[0,1)})^2$, the agents have a lot of bids with a high utility, and it as shown in Fig. 8.

4.2 *Experiments for Cooperativeness Measures*

We perform bilateral negotiations and three-party negotiations in a round-robin and compare the correlation coefficients between the social welfare of these negotiations and each cooperativeness measure. In three-party negotiations with our domains, we only use profiles 1-10 because of the computational time limitation. The details of the negotiation settings in this experiment are as follows:

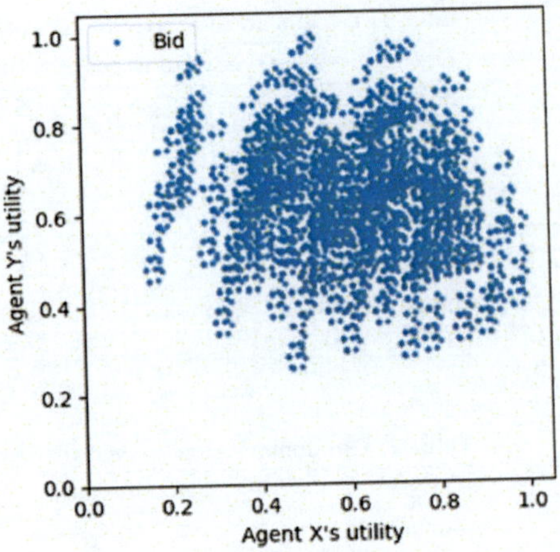

Fig. 6 Utility diagram of the domain_5a

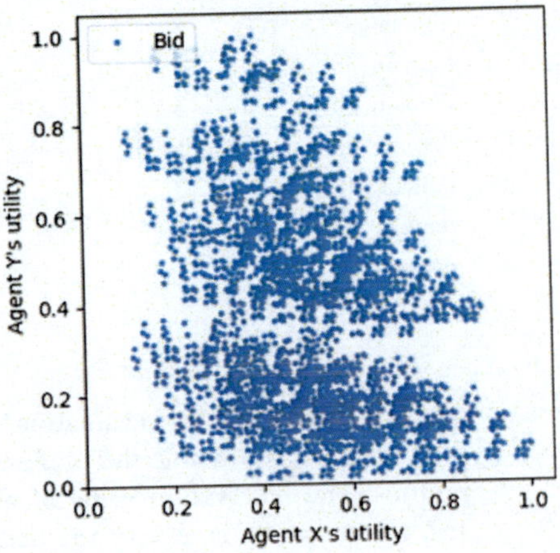

Fig. 7 Utility diagram of the domain_5b

- Tournament setting: round-robin tournaments
- Number of agents in negotiations: 2 or 3
- Agents: top three agents in the individual utility categories in ANAC 2016 (Caduceus, YXAgent, ParsCat)
- Negotiation domains: All domains in Tables 1 and 2. In three-party negotiations, we only use profiles 1–10 in our domains.
- Negotiation protocol: SAOP
- Deadline: 10,000 rounds.

Table 3 shows the correlation coefficient between social welfare and each cooperativeness measure in bilateral negotiations. Since domain1 has only one bid which is more than the reservation values, we cannot calculate the correlation coefficient of CCC_{RV} in domain1. Furthermore, Table 4 shows the correlation coefficient between

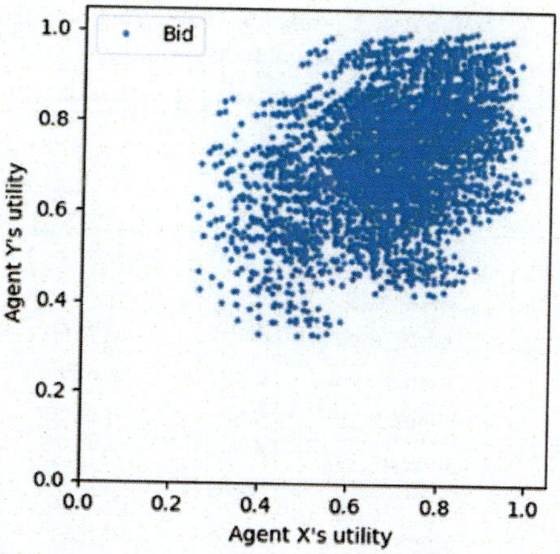

Fig. 8 Utility diagram of the domain_5c

Table 3 Correlation coefficients between social welfare and each cooperativeness measure in bilateral negotiations

Domain	CHI	CHI_{RV}	CCC	CCC_{RV}	MOL_{ALL}	MOL_{RV}	MOL_{PO}	MOL_{RVPO}
domain1	0.801	0.910	0.952	–	−0.806	0.262	−0.510	0.262
domain2	0.946	0.932	0.685	0.835	−0.643	−0.619	−0.924	−0.861
domain4	0.964	0.964	0.919	0.919	−0.872	−0.872	−0.956	−0.956
domain8	0.943	0.943	0.803	0.803	−0.879	−0.879	−0.890	−0.890
domain16	0.968	0.939	0.881	0.743	−0.874	0.733	−0.952	−0.031
party_domain	0.920	0.920	0.648	0.648	−0.562	−0.562	−0.834	−0.834
domain_3a	0.931	0.876	0.629	0.523	−0.584	−0.180	−0.670	−0.296
domain_3b	0.932	0.909	0.768	0.583	−0.532	0.181	−0.740	−0.038
domain_3c	0.924	0.868	0.662	0.721	−0.373	−0.607	−0.567	−0.615
domain_5c	0.942	0.912	0.629	0.698	−0.367	0.159	−0.665	−0.576
domain_5b	0.946	0.914	0.663	0.643	−0.392	0.541	−0.693	−0.077
domain_5c	0.886	0.872	0.604	0.650	−0.264	−0.213	−0.678	−0.663
domain_7a	0.929	0.911	0.702	0.767	−0.532	−0.204	−0.704	−0.658
domain_7b	0.957	0.947	0.731	0.755	−0.536	0.585	−0.796	−0.255
domain_7c	0.845	0.840	0.572	0.605	−0.329	−0.271	−0.615	−0.605
Overall	0.930	0.887	0.538	0.538	−0.500	−0.147	−0.770	−0.535

social welfare and each cooperativeness measure in three-party negotiations. We did not evaluate CCC in the three-party negotiation because it cannot be applied to three-party negotiations. In addition, the correlation coefficient in domain16 cannot be calculated because all negotiations in the domain failed. This result is caused by extremely high reservation values of domain16.

CHI has a significantly stronger correlation with social welfare than other cooperativeness measures except for CHI_{RV} ($p < 0.05$, Welch's t-test). In some domains,

Table 4 Correlation coefficients between social welfare and each cooperativeness measure in three-party negotiations

Domain	CHI	CHI_{RV}	MOL_{ALL}	MOL_{RV}	MOL_{PO}	MOL_{RVPO}
domain1	0.752	0.915	−0.642	0.678	−0.459	0.678
domain2	0.859	0.808	−0.792	−0.840	−0.886	−0.947
domain4	0.836	0.836	−0.710	−0.710	−0.764	−0.764
domain8	0.823	0.823	−0.763	−0.763	−0.795	−0.795
domain16	–	–	–	–	–	–
party_domain	0.755	0.755	−0.355	−0.355	−0.354	−0.354
domain_3a	0.903	0.879	−0.566	−0.540	−0.398	−0.460
domain_3b	0.847	0.804	−0.484	0.333	−0.476	0.476
domain_3c	0.739	0.750	−0.299	−0.369	−0.375	−0.338
domain_5c	0.908	0.936	−0.466	0.636	−0.645	−0.065
domain_5b	0.895	0.898	−0.262	0.624	−0.508	0.697
domain_5c	0.801	0.782	−0.103	−0.024	−0.423	−0.474
domain_7a	0.913	0.910	−0.567	0.170	−0.718	−0.473
domain_7b	0.938	0.950	−0.532	0.849	−0.768	0.681
domain_7c	0.700	0.696	−0.209	−0.133	−0.459	−0.438
Overall	0.869	0.805	−0.534	−0.179	−0.700	−0.272

other methods have a stronger correlation than CHI; however, CHI has the strongest correlation in almost all domains. This result shows that CHI can accurately evaluate domains using the hypervolume indicator in various situations. The overall correlation of MOL_{PO} is weaker than CHI because MOL focuses only on fairness. Since MOL does not consider the social welfare, MOL evaluates a pair that has a lot of fair bids with low social welfare as a cooperative pair. The results of MOL_{ALL} and MOL_{PO} demonstrate the effectiveness of evaluating only Pareto optimal bids in MOL. In most domains, a correlation of MOL is weaker than the overall correlation. It is shown that MOL is inadequate to compare pairs or groups with similar comparable cooperativeness. The overall correlation of CCC is weaker than those of CHI and MOL_{PO}. This result is attributed to situations wherein agents can cooperate without similar preferences. An instance is a situation wherein both agents emphasize different issues. Furthermore, the evaluation of only the bids that are more than the reservation values is inefficient to measure their cooperativeness. In particular, MOL_{RV} and MOL_{RVPO} show this characteristic.

In most cooperativeness measures, the correlations in domains 3b, 5b, and 7b are larger than the ones in domains 3c, 5c, and 7c because the variance of each pair's social welfare between the domains is different. In domains 3c, 5c, and 7c, the variances of the social welfare are small because the social welfare of many pairs is almost identical to the maximum value. In contrast, in domains 3b, 5b, and 7b, the variances of the social welfares are large because the cooperative and competitive pairs are mixed.

4.3 Experiments for Matching Method

We perform bilateral negotiations in the round-robin and compare the total social welfare between the matchings using the proposed method and theoretical optimal solutions. First, we perform bilateral negotiations with each pair to obtain the social welfare. We use only our domains in the experiments because the number of profiles of GENIUS domains was too small. We prepare three versions with different preference profiles using the same settings for each domain. The details of the negotiation settings in this experiment are as follows:

- Tournament setting: round-robin tournament
- Number of agents: 2
- Agents: Top three agents in the individual utility categories in ANAC 2016 (Caduceus, YXAgent, ParsCat)
- Negotiation domains: All domains in Table 2; each domain has three versions with different preference profiles.
- Negotiation protocol: SAOP
- Deadline: 10,000 rounds.

Then, we exhaustively search for the theoretical optimal solution in each domain based on the negotiation results. In this paper, the theoretical optimal solution is defined as maximizing the sum of the social welfare of each matching pair. Finally, we compare the matchings using the proposed method with the theoretical optimal solutions. R is the ratio of the total social welfare between the matchings using our method and the theoretical optimal matchings, and it is defined as follows:

$$R = \frac{\sum_{p \in m} \mathrm{SW}(p)}{\sum_{p^* \in m^*} \mathrm{SW}(p^*)},$$

where m is the matching of the proposed method, m^* is the theoretical optimal solution, and $SW(p)$ is the social welfare of pair p in this experiment.

Table 5 shows the ratios of total social welfare between matchings from our method and theoretical optimal solutions. A column shows the cooperativeness measure that is used in the matching method, where RAND represents the results of matchings generated randomly.

The matching method based on CHI has a significantly higher R value than methods based on other cooperativeness measures except for $\mathrm{CHI_{RV}}$ ($p < 0.05$, Welch's t-test). The reason why the method based on CHI outperforms the other methods is that CHI has a strong correlation with social welfare than other cooperativeness measures. Furthermore, our method based on CHI, $\mathrm{CHI_{RV}}$, CCC, $\mathrm{MOL_{ALL}}$, $\mathrm{MOL_{PO}}$ and $\mathrm{MOL_{RVPO}}$ significantly outperforms RAND ($p < 0.05$, Welch's t-test). It demonstrates that our method based on an appropriate cooperativeness measure can pair agents which shared mutual interests.

Table 5 Ratios of the total social welfare between the matchings from our method and the theoretical optimal matchings (*R*)

Domain	CHI	CHI_{RV}	CCC	CCC_{RV}	MOL_{ALL}	MOL_{RV}	MOL_{PO}	MOL_{RVPO}	RAND
domain_3a	0.997	0.997	0.934	0.905	0.987	0.940	0.944	0.944	0.877
domain_3b	0.998	0.991	0.936	0.639	0.960	0.801	0.945	0.803	0.802
domain_3c	0.996	0.996	0.966	0.969	0.998	0.997	0.967	0.971	0.932
domain_5a	0.993	0.993	0.952	0.900	0.986	0.990	0.965	0.979	0.906
domain_5b	0.999	0.997	0.956	0.689	0.984	0.874	0.977	0.964	0.813
domain_5c	0.995	0.995	0.972	0.981	0.985	0.985	0.984	0.985	0.952
domain_7a	0.993	0.993	0.980	0.969	0.990	0.990	0.979	0.986	0.943
domain_7b	0.993	0.993	0.977	0.787	0.986	0.950	0.982	0.983	0.885
domain_7c	0.994	0.994	0.972	0.973	0.990	0.990	0.971	0.978	0.950
Average	0.995	0.994	0.961	0.868	0.985	0.946	0.968	0.955	0.896

5 Related Works

Brzostowski and Kowalczyk proposed an approach of selecting opponents according to case-based reasoning [6]. Based on the history of past negotiations, this approach predicts an agent's behavior in the form of a possibility distribution. The constructed distribution is used to calculate the expected utility to evaluate the chance of reaching an agreement. This approach requires the history of past negotiations to evaluate the relationship between agents. On the other hand, our method does not require the negotiation history; therefore, it can be used in completely new situations.

The International Automated Negotiating Agent Competition (ANAC) has been held annually since the first competition in 2010 [1, 2, 4]. ANAC was organized to promote research into automated negotiation, particularly multi-issue closed negotiation. The followings are the goals of ANAC: (1) providing a motivation for the development of negotiation protocols and negotiation strategies; (2) collecting and developing a benchmark of negotiation scenarios, protocols, and strategies; (3) developing a common set of tools and criteria for their evaluation; and (4) setting the research agenda for automated negotiation. The negotiation platform of ANAC is GENIUS, which stands for a General Environment for Negotiation with Intelligent multi-propose Usage Simulation [15].

Shinohara and Fujita proposed a negotiation protocol that addresses the fairness of revealing each agent's private information [20]. The protocol reveals private information of the agent who reveals less private information by increasing the number of its offers. Zhang et al. proposed a hierarchical negotiation protocol for domains comprising a large number of interdependent issues [5]. The protocol improves scalability by arranging the issues hierarchically based on the preference of the agents. Hoz et al. proposed a method of Wi-Fi channel assignment based on automated negotiation [10]. The method improves the throughput using the channel negotiation by provider agents that manage some access points. In addition, the automated negotiations are applied to human-agent negotiations [18] and game-playing [11].

Most existing works related to automated negotiations assume that the opponents are determined beforehand. However, selecting opponents is one of the most important elements in real-life negotiations. Therefore, we focused on measuring cooperativeness and agent matchings in this paper.

6 Conclusions and Future Work

In this paper, we proposed the cooperativeness measure based on the hypervolume indicator (CHI) as a measure of the relations between agents. CHI is based on the hypervolume indicator that is the measure to evaluate the results of evolutionary multi-objective optimization algorithms. CHI is defined as the hypervolume of the negotiation space, which is a Pareto dominated by the set of all the bids. We demonstrated that the effectiveness of CHI by the tournaments among the top three agents of ANAC 2016 in GENIUS. In this experiment, we compared three cooperativeness measures: CHI, CCC, and MOL. CCC is the cooperativeness measure based on the correlations between agents' utilities, and MOL is the cooperativeness measure based on the fairness of the agents' utilities. The correlation with the social welfare of CHI outperformed the other methods. The experimental results also demonstrated that CHI effectively evaluated the cooperativeness of the agents.

Furthermore, we proposed a matching method based on the cooperativeness measure for concurrent negotiations. In our proposed method, the mediator collected the preference information of all the agents and calculated the cooperativeness in order to find cooperative pairs. Our method found the matching with the largest sum of cooperativeness. We compared the total social welfare between the matching of our method and the theoretical optimal solution in order to evaluate our method. In this paper, the theoretical optimal matching was defined as the matching with the largest sum of social welfare. Our method using CHI provided the highest average optimization ratio of the total social welfare. In all domains, the optimization ratios of our method based on CHI were extremely high. Therefore, we demonstrated that our matching method based on CHI can find the cooperative matchings.

One important future work is to consider the privacy of agents. Generally, automated negotiation assumes closed negotiations, which private the utility and strategy of agents during the negotiation. However, our approach requires opening the utility of the agent to the mediator. To solve this, estimating the cooperativeness from the limited offers in the negotiation is necessary.

References

1. R. Aydogan, T. Baarslag, K. Fujita, K. Hindriks, T. Ito, C. Jonker, in The Seventh International Automated Negotiating Agents Competition (ANAC2016). http://web.tuat.ac.jp/katfuji/ANAC2016/ (2016)
2. R. Aydogan, T. Baarslag, K. Fujita, T. Ito, D. de Jonge, C. Jonker, J. Mell, in The Eighth International Automated Negotiating Agents Competition (ANAC2017). http://web.tuat.ac.jp/katfuji/ANAC2017/ (2017)
3. R. Aydoğan, D. Festen, K.V. Hindriks, C.M. Jonker, *Alternating Offers Protocols for Multilateral Negotiation* (Springer International Publishing, Cham, 2017), pp. 153–167
4. T. Baarslag, R. Aydogan, K.V. Hindriks, K. Fujita, T. Ito, C.M. Jonker, The automated negotiating agents competition, 2010–2015. AI Mag. **36**(4), 115–118 (2015)
5. C.E. Brodley, P. Stone, ed., in *Proceedings of the Twenty-Eighth AAAI Conference on Artificial Intelligence*, July 27–31, 2014 (. AAAI Press, Québec City, Canada, 2014)
6. J. Brzostowski, R. Kowalczyk, in On possibilistic case-based reasoning for selecting partners for multi-attribute agent negotiation, in *Proceedings of the Fourth International Joint Conference on Autonomous Agents and Multiagent Systems (AAMAS '05)*, New York, NY, US. ACM (2005), pp. 273–279
7. S. Shaheen Fatima, N.R. Jennings, M.J. Wooldridge, Multi-issue negotiation with deadlines, in *CoRR*, abs/1110.2765 (2011)
8. S. Fatima, S. Kraus, M. Wooldridge, *Principles of Automated Negotiation* (Cambridge University Press, Cambridge, 2014)
9. Z. Galil, Efficient algorithms for finding maximum matching in graphs. ACM Comput. Surv. **18**(1), 23–38 (1986)
10. E. de la Hoz, I. Marsá-Maestre, J.M. Giménez-Guzmán, D. Orden, M. Klein, Multi-agent nonlinear negotiation for wi-fi channel assignment. In Larson et al. [14], pp. 1035–1043
11. D. de Jonge, D. Zhang, Automated negotiations for general game playing. In Larson et al. [14], pp. 371–379
12. S. Kraus, *Strategic Negotiation in Multiagent Environments, Intelligent robots and autonomous agents* (MIT Press, Cambridge, 2001)
13. S. Kraus, J. Wilkenfeld, G. Zlotkin, Multiagent negotiation under time constraints. Artif. Intell. **75**(2), 297–345 (1995)
14. K. Larson, M. Winikoff, S. Das, E.H. Durfee, ed., in *Proceedings of the 16th Conference on Autonomous Agents and MultiAgent Systems (AAMAS 2017)*, São Paulo, Brazil, May 8–12, 2017. ACM (2017)
15. R. Lin, S. Kraus, T. Baarslag, D. Tykhonov, K. Hindriks, C.M. Jonker, Genius: an integrated environment for supporting the design of generic automated negotiators. Comput. Intell. **30**(1), 48–70 (2014)
16. I. Marsá-Maestre, M. Klein, E. de la Hoz, M.A. López-Carmona, Negowiki: a set of community tools for the consistent comparison of negotiation approaches. In Kinny D., Hsu J.Y., Governatori G., Ghose A.K. (eds) Agents in Principle, Agents in Practice. PRIMA 2011. Lect. Notes Comput. Sci. vol 7047. Springer, Berlin, Heidelberg, pp. 424–435
17. I. Marsá-Maestre, M. Klein, C.M. Jonker, R. Aydogan, From problems to protocols: towards a negotiation handbook. Decis. Supp. Syst. **60**, 39–54 (2014)
18. J. Mell, J. Gratch, IAGO: interactive arbitration guide online (demonstration). In Jonker et al. *Proceedings of the 15th International Conference on Autonomous Agents and Multiagent Systems* (AAMAS 2016), pp. 1510–1512
19. A. Rubinstein, Perfect equilibrium in a bargaining model. Econometrica J. Econom. Soc. pp. 97–109 (1982)
20. H. Shinohara, K. Fujita, Alternating offers protocol considering fair privacy for multilateral closed negotiation. In An B., Bazzan A., Leite J., Villata S., van der Torre L. (eds) PRIMA 2017: Principles and Practice of Multi-Agent Systems. PRIMA 2017. Lect. Notes Comput. Sci, vol 10621. Springer, Cham, pp. 533–541

21. Y. Shoham, K. Leyton-Brown, *Multiagent Systems—Algorithmic, Game-Theoretic, and Logical Foundations* (Cambridge University Press, Cambridge, 2009)
22. T. Toyama, A. Moustada, T. Ito, On measuring the opposition level amongst intelligent agents in multi-issue closed negotiation scenarios, in *12th International Conference on Knowledge, Information and Creativity Support Systems (KICSS 2017)* (2017)
23. E. Zitzler, L. Thiele, Multiobjective optimization using evolutionary algorithms – A comparative case study. In: Eiben A.E., Bäck T., Schoenauer M., Schwefel HP. (eds) *Parallel Problem Solving from Nature – PPSN V. PPSN 1998.* Lect. Notes Comput. Sci. vol 1498. Springer, Berlin, Heidelberg (1998)